Beginning jOOQ

Learn to Write Efficient and Effective Java-Based SQL Database Operations

Tayo Koleoso

Apress®

Beginning jOOQ: Learn to Write Efficient and Effective Java-Based SQL Database Operations

Tayo Koleoso
Silver Spring, MD, USA

ISBN-13 (pbk): 978-1-4842-7430-9 ISBN-13 (electronic): 978-1-4842-7431-6
https://doi.org/10.1007/978-1-4842-7431-6

Managing Director, Apress Media LLC: Welmoed Spahr
Acquisitions Editor: Steve Anglin
Development Editor: Matthew Moodie
Coordinating Editor: Mark Powers

Cover designed by eStudioCalamar

Cover image by Josh Rose on Unsplash (www.unsplash.com)

Distributed to the book trade worldwide by Apress Media, LLC, 1 New York Plaza, New York, NY 10004, U.S.A. Phone 1-800-SPRINGER, fax (201) 348-4505, e-mail orders-ny@springer-sbm.com, or visit www.springeronline.com. Apress Media, LLC is a California LLC and the sole member (owner) is Springer Science + Business Media Finance Inc (SSBM Finance Inc). SSBM Finance Inc is a **Delaware** corporation.

For information on translations, please e-mail booktranslations@springernature.com; for reprint, paperback, or audio rights, please e-mail bookpermissions@springernature.com.

Apress titles may be purchased in bulk for academic, corporate, or promotional use. eBook versions and licenses are also available for most titles. For more information, reference our Print and eBook Bulk Sales web page at http://www.apress.com/bulk-sales.

Any source code or other supplementary material referenced by the author in this book is available to readers on GitHub via the book's product page, located at www.apress.com/9781484274309. For more detailed information, please visit http://www.apress.com/source-code.

Printed on acid-free paper

Table of Contents

About the Author

Tayo Koleoso is the Founder and CEO of LettuceWork (`www.lettucework.io`), the platform dedicated to engineering culture. He created the Better Managed Development method for building and sustaining an effective product engineering culture. He's a lifelong learner, engineer, and engineering leader committed to building people and software in a healthy, sustainable, and effective ecosystem. Outside of tech, comedy is the only thing he consumes in large quantity. King of the Hill, Peep Show and 30 Rock are his comfort telly, *I tell you what.*

He got his start in software engineering as a teenage database programmer with Oracle 8i. The jOOQ platform is therefore a natural fit and a return to his roots: his love affair with SQL.

About the Technical Reviewer

Vishwesh Ravi Shrimali graduated in 2018 from BITS Pilani, where he studied mechanical engineering. Currently, he is working at Mercedes Benz Research and Development India Pvt. Ltd. as an ADAS Engineer. He has also co-authored *Machine Learning for OpenCV 4* (Second Edition), *The Computer Vision Workshop*, and *Data Science for Marketing Analytics* (Second Edition) by Packt. When he is not writing blogs or working on projects, he likes to go on long walks or play his acoustic guitar.

CHAPTER 1

Welcome to jOOQ

I got my start in software engineering (and really, serious computer business) at 15 years old, with Oracle 8i SQL. Yes, I've been an old man from a young age, technologically speaking. Playing with SQL* Plus, trying (and failing) my first Oracle SQL certification exam, before I even started university, taught me the value of getting SQL right. Don't take it from me, take it from this chap:

> *I was a data access purist: I like my DAOs chilled, my PreparedStatements prepared, and my SQL handwritten with the care and tenderness of a lover... The world moved on to Hibernate, Java Persistence API (JPA), and everything in between... I still believe in raw SQL – a well-crafted SQL statement will outperform Object-Relational Mapping (ORM).*
>
> —A tall, dark, dashing young and cool man, with flowing locks of jet black hair and piercing brown eyes[1]

That tall drink of SQL? Probably me; I don't know. The point is I deeply appreciate Structured Query Language (SQL) and all it has to offer. The industry's been going gaga about NoSQL because it's "easy to use" and it "scales quickly," but the fact of the matter is that SQL is still the undisputed king of Online Analytical Processing (OLAP). When you want sanity and integrity in your data, SQL is there. When you want (most of[2]) the

[1] Editor's note: Oh brother. Here we go again.

[2] I say "most of" here because different Relational Database Management Systems provide varying degrees of ACID strength.

T. Koleoso, *Beginning jOOQ*, https://doi.org/10.1007/978-1-4842-7431-6_1

guarantees of reliable transaction handling (à la ACID), you're still going to need solid SQL in your stack. Not for nothing, database stored procedures will typically outperform application-layer (e.g., Java, .Net) processing in many cases. In the words of the late, great Thanos: "SQL is inevitable. It's in the interests of your application's scalability and correctness to get it right."

Unfortunately, SQL gets very short shrift from devs nowadays. The database is just another "black box" that we're supposed to yell commands at, so it yields some data and then we move on. It's not until our queries progressively degrade due to preventable problems; our schema is an incoherent mess after two versions of our applications; SQL injection attacks expose our weaknesses; the application chokes on queries returning more than a few hundred rows. One of the dark sides of SQL is that you're not likely to realize that your SQL query is returning incorrect or incomplete data at first glance. You ran a query, it returned *some* queries, and that's that, right? Yikes.

This book isn't about the fundamentals of SQL. Or even the joys of SQL per se (there are many). This book is about taking a different look at handling SQL work in Java.

Database Operations in Java: The Good Parts

Your options for handling SQL data in the Java world are fairly straightforward:

1. JDBC (Java Database Connectivity): JDBC is the most fundamental API supporting Relational Database Management System (RDBMS) access. It provides

    - Connection management
    - Direct SQL statement control

- Stored procedure and function execution
- Mostly SQL injection safe componentry
- Transaction management

Save for one or two JakartaEE specifications, pretty much everything else RDBMS related in the Java ecosystem is based on JDBC. Because of JDBC, we can then have...

2. Mapping Frameworks: Yes, I'm talking about Object-Relational Mapping (ORM) frameworks like Hibernate, MyBatis, and EclipseLink. These are *très* convenient frameworks, based on the premise that developers don't want to spend any time...developing SQL or other database-related constructs. Neato. With these mapping frameworks, you get to define some classes, slap some annotations on them and the framework:

    - Maps your java classes (the object model) to your database tables (the domain model). This mapping is used to convert query results into java objects, known as *entities*. These entities are *managed objects* - like a concierge service - changes to the entities in memory are tracked and persisted by the ORM runtime.
    - Allows you to declaratively model the relationships between your tables in your RDBMS, using java object relationships (is-a, has-a type stuff). An absolute cornucopia of annotations supports this feature of ORMs.

- Completely spares you of any details related to the SQL involved in all this magic. It just works.
- Provides declarative transaction handling - with more annotations.
- Provides an *additional* query language, Hibernate Query Language (HQL), that introduces an object-oriented flavor to the mix. This way, you can abandon SQL altogether (!) and just speak fluent OOP all the time!

Most ORM providers offer some form of caching of the results of database queries. The goal here is to save the travel time to the database for subsequent trips to the database. So that when one user loads some data once, if they request the same rows, it's already in memory.

Then we have the Java Persistence API (JPA). This is a JakartaEE specification that attempts to standardize the usage and behavior of ORMs in the Java platform. The various providers (Hibernate, etc.) implement this specification to varying degrees. They each also have implementation-specific syntactic sugar that isn't supported by the API. The API still allows you to write your raw SQL if you like, and the results can still be managed objects. Pretty neat.

In addition to all this, a framework like Spring offers the `JdbcTemplate` as a wrapper around JDBC proper. SQL in the Java system is just one raging party of convenience. Nice!

Database Operations in Java: The...Not So Good Parts

Ask yourself this: why aren't JavaServer Pages (JSP) and JavaServer Faces (JSF) as wildly popular as, say, React.js or Vue.js, when front-end development is concerned? A lot of Java-based organizations are happy to have Java or Kotlin Spring Boot back ends, but fronted by *not Java*. Because when you care about performance and resource efficiency in a domain like the browser, nothing beats raw JavaScript.

And I say this as someone that's spent a fair bit of time teaching about JSF and answering questions on StackOverflow. Don't get me wrong: JSF is super convenient. Heck, that's why I got into the business of JSF in the first place: a cheap, convenient, and practical way to belch out markup and scripting into a web page. But when no one's watching, *I know*. I know that raw JavaScript is still where it's at. If you want to make your browser dance, deal with the quirks and nuances of individual browsers, you turn to the language invented for browsers. These ~~hips~~ stylesheets don't lie.

Yet here we are, where many have decided that SQL should take a backseat when interacting with databases. Park the language built for the platform in the garage; Java, the language of kings, is preferable. In many scenarios, it isn't. Here are some reasons why:

1. JPA isn't aware of what type of database you're using, which is a shame, when you consider that there are specific quirks, features, and limitations of individual databases, for example:

    - MySQL doesn't support the INTERSECT or EXCEPT set operations; FULL JOIN is also off the menu. You wouldn't know until you tried to use it and your operation chokes.

    - JPA doesn't know what to do with nulls in the `ORDER BY` clause; there's also no support for the `ORDER BY NULLS FIRST` clause from standard SQL.[3] You're on your own here.

[3] Coming soon with Hibernate 6: `https://docs.jboss.org/hibernate/orm/6.0/userguide/html_single/Hibernate_User_Guide.html#hql-order-by`

- JPA doesn't deal well with the `IN` clause in some scenarios:
  - When you want to take advantage of query plan caching
  - When there are nulls in the list of parameters passed to the `IN` clause
- PostgreSQL supports a massive array of data types that are hyper-specific and hyper-optimized to some use cases. There's a wide assortment of data types you can leverage in this RDBMS that you will have to do a bunch of extra work to support with `UserTypes` in Hibernate.
- Many of the mainstream database providers (Oracle, PostgreSQL, and MySQL at least) provide document storage and SQL querying – that's right, you can save your JSON documents in these databases, query, and navigate *inside* the documents with SQL. Basically combine NoSQL and SQL in the same box. Some benchmarks have shown the performance to be comparable to the likes of MongoDB up to certain scales. These aren't your grandmother's RDBMSes.

"But I want to make my application portable." Your enterprise has spent borderline sinful sums of money on an Oracle license, but you're going to use like 5% of its capabilities, like a really fancy Excel spreadsheet?

2. Even with native query capabilities, neither JPA nor Hibernate will save you from yourself. Your raw SQL is still open to SQL injection if you make the right mistakes. Your SQL could still be incorrect, and you won't find out until you try to execute the native query. Java Persistence Query Language (JPQL) and Hibernate Query Language (HQL) aren't going to save you either. You won't find out your query syntax is broken or incorrect until you try to run it. And if you accidentally make changes to a managed JPA entity, it's going to be committed to the database the first chance it gets.

3. Remember the caching that Hibernate and other tools will do for you by default? Guess whose RAM is slowly being devoured? Go on, guess. You might be surprised to find out that every entity retrieved and managed by a single hibernate session is cached – for just that hibernate session – so that in a large enterprise application with any number of concurrent users, they're all liable to hold copies of exactly the same data in RAM of the application server. Imagine how thirsty your application will get at scale!

4. Can you confidently say you know what's going on inside of Hibernate or EclipseLink? Have you tried to look at the actual queries being generated by Hibernate? You might be in for a lot of disappointment. There are many scenarios where Hibernate is simply wasteful with database trips that are hidden from view:

- Batch inserts and updates aren't enabled by default, and you're going to do a bit of work to fully support both.
- Even more specifically, using `GenerationType.IDENTITY` with PostgreSQL and some others, Hibernate will still ignore any batching directives.[4]

5. The challenges of an ORM really get in your face when you need to scale. A couple rows, maybe a couple hundred, and you could skate by. Larger result sets, on the other hand, are sometimes not practical (see the previous discussion: loading all retrieved entities into memory). You could struggle to handle an increase in query volume. JPA 2.2 introduced support for more efficient streaming from the database; but again different kinds of databases handle this feature differently. So that despite your best efforts, MySQL and PostgreSQL could still very well retrieve the entire `ResultSet`,[5] ignoring your JPA 2.2 expectations of efficient results streaming.

6. Fundamentally, ORM as a concept is at odds with relational data representations. The technical name for it is the Object-Relational Impedance Mismatch. Don't take it from me; ask the nerds at Hibernate themselves[6]:

---

[4] https://vladmihalcea.com/postgresql-serial-column-hibernate-identity/

[5] https://vladmihalcea.com/whats-new-in-jpa-2-2-stream-the-result-of-a-query-execution/

[6] http://hibernate.org/orm/what-is-an-orm/

> The way you access data in Java is fundamentally different than (sic) the way you do it in a relational database. In Java, you navigate from one association to another walking the object network. This is not an efficient way of retrieving data from a relational database. You typically want to minimize the number of SQL queries and thus load several entities via JOINs and select the targeted entities before you start walking the object network.

Point is that past a certain point, you're either going to be dealing with a mess of annotations and a hopelessly complex entity class graph, or you're going to need to roll up your sleeves and get SQL-y.

You can't avoid writing SQL if you're going to do data access correctly in Java. This isn't a dig at the gang over at Hibernate; that framework is a dang miracle for many use cases. But we're not here to talk about Hibernate, are we?

You Have Got to Be jOOQing

Yes, I'm here to give you the good news of the jOOQ framework. First things first: jOOQ is not a complete replacement for Hibernate, JPA, or anything in that realm. JPA delivers on its goals of standardizing most of RDBMS access for Java developers. Hibernate is hella powerful and convenient; particularly for write operations, you can ORM to your heart's content and delight. I mean, what can beat the convenience of simply updating a field of a java object and `persist`-ing or `flush`-ing?

For adult-tier, large-scale SQL data wrestling? Where there are a *lot* of read operations? You'll need the big guns. jOOQ is as big a gun as you'll get in this business of ours. Don't take it from me, take it from an actual ORM expert:

How about this chap:

> *...But while abstracting the SQL write statements is a doable task, when it comes to reading data, nothing can beat native SQL...native querying is unavoidable on almost any enterprise system...While you can fire native queries from JPA, there's no support for dynamic native query building. jOOQ allows you to build typesafe dynamic native queries, strengthening your application against SQL injection attacks. jOOQ can be integrated with JPA, as I already proven on my blog, and the JPA-jOOQ combo can provide a solid data access stack.*
>
> —Vlad Mihalcea[7]

Vlad was a contributor to the Hibernate platform and continues to support his own query performance optimizer, and he's an authority on the relationship between Java and relational databases. Cheers to that.

If you think of JPA and its implementations as being *too* Object-Oriented Programming (OOP) friendly, jOOQ seeks to bring a balance to the force. Providing strong awareness of the nuances of safe, effective, and efficient SQL handling, all the while maintaining a solid grip on the object-oriented side of things. So what's in the jOOQ box?

jOOQ Feature Tour

Alright, gather round everyone[8]! Let me sing you the song of ~~my people~~ jOOQ. Ohhh, this is a tale of...

[7] `https://blog.jooq.org/2015/04/14/jooq-tuesdays-vlad-mihalcea-gives-deep-insight-into-sql-and-hibernate/`

[8] Editor's note: How many people do you think are concurrently reading a single copy of your book?

Database Aware

jOOQ is unapologetically RDBMS vendor aware.[9] Unlike *those other guys*, jOOQ provides API kits built specifically for a wide range of major database vendors. This way, when you're using jOOQ tools and APIs, you can be sure you're idiomatically correct; the syntax and semantics of your code is transparently optimized and tightly mated with your database of choice. And the delightful part of this is that it doesn't translate to a tight coupling to your Java code. Because jOOQ skews closer to the database, your Java code can still remain technically database agnostic; your jOOQ library implementation will quietly handle the finer details and nuances of your specific database of choice. You can then quite easily flip a configuration in jOOQ to support another, should you choose to switch databases.

Code Generation

jOOQ is *all* about convenience when it comes to the developer experience. Observe.

Starting with any one of the following:

- Existing database schema
- Database table
- Stored procedure or function
- Sequence
- JPA entity
- XML file

[9] `www.jooq.org/download/support-matrix`

- Plaintext file containing SQL
- Liquibase YAML, XML, or JSON file
- Custom data types

jOOQ will generate java classes, JPA-annotated entity classes, interfaces, or even XML that correctly maps what is represented in the source material. If you're starting with stored procedures or functions, jOOQ will generate "routines": classes that represent that database code, making it as convenient as physically possible to use what's already available.

Type Safety

All the entities, data, and fields you'll be working with are typesafe – guaranteed correct. The fields from your database tables, bind variables from your stored procedures and functions, etc. are translated to the closest match available in Java.

Domain-Specific Language

jOOQ provides a DSL that mirrors the specific SQL dialect that's implemented by your choice of database vendor. Even beyond that, the library integrates the quirks and non-standard features of the supported database vendors. You know the ones, the *bad boy* databases (looking at you, *Oracle*). This compile-time checking of your SQL syntax means that you'll have to go out of your way to write broken or otherwise unsupported SQL. No more waiting till runtime to find out your SQL doesn't compile!

Tooling Support

You also get a robust toolkit to support a lot of developer convenience and safe usage of the library:

- Maven: jOOQ offers configurable Maven plugins and tools to support a variety of tasks around the code generation feature.
- Command Line: Also supports the powerful code generation feature of jOOQ. Additionally, there's a Schema Diff feature that helps compare two schemata and output the difference.
- Flyway and Liquibase: You can use the two most popular DML and DDL version control libraries with jOOQ.
- Raw SQL Vendor Safety: Should you opt to not use the SQL DSL, instead choosing raw SQL, you can use jOOQ's `Parser` anywhere in your code to ensure that the SQL you're writing is correct and valid for the selected database.
- SQL Disabler: The `PlainSQLChecker` allows you to disable support for raw SQL anywhere in your project. Drop it in your Maven `POM.xml` and it'll disallow compilation of any raw SQL in your project. This way, you can guarantee that anyone that's working on your project or codebase will have to use jOOQ-generated code and jOOQ's fluent API; eliminating the probability of incorrect SQL creeping into the code.

JVM Languages

We get to take the jOOQ box on tour! You can use jOOQ with

- Kotlin
- Scala
- Groovy

This has been a 50,000 ft. flyover of jOOQ and what it brings to the table. I want to stress here that jOOQ isn't a replacement for ORM – the right tool for the right job, etc. Where Hibernate, etc. don't give you the coverage you need, that's where jOOQ swings into action.

So! *Whaddya think so far?* Pretty neat, huh? I'll just catch my breath for a moment here, and see you in the next chapter!

CHAPTER 2

Getting Started with jOOQ

Let's take this thing for spin! But first, you should get to know these classes and interfaces; they're at the heart of 75% of what you'll be doing in jOOQ:

1. `org.jooq.impl.DSL`

   This is the sturdy, wise grandparent of most of jOOQ's functionality. With this class, you can

   - Manipulate parts of a strongly typed SQL query in a typesafe manner.
   - Access database functions (`concat`, `coalesce`, etc.) as if they were java methods.
   - Specify database dialects. This is for when you need to perform operations that are unique to your datastore (Oracle, Amazon Redshift, PostgreSQL, etc.).
   - Carry out Database Definition Language (DDL) operations like `ALTER`, `CREATE`, `DROP`, etc., all in a typesafe manner.
   - Perform more mundane tasks like
     - Constructing plaintext SQL
     - Configuring the database connections

T. Koleoso, *Beginning jOOQ*, https://doi.org/10.1007/978-1-4842-7431-6_2

Almost all of its functionality are static methods in the class, so you'll typically just need to perform an `import static org.jooq.impl.DSL.*` to use its features.

2. `org.jooq.DSLContext`

   Where `DSL` offers almost all its functionality in static methods without state, DSLContext is more object oriented. There's some overlap between this component and the `DSL` class, but there's a fundamental difference. It's really in the name DSL: Domain-Specific Language **Context**. Keeping state in a context object provides runtime capabilities that you're not going to get from `DSL` – comes in handy when your SQL operation is a multi-step one or is part of a longer process. Overall, `DSLContext` offers

   - Fluent programming style
   - Stateful components
   - Better integration with dependency injection frameworks like Spring (more on that later)

   The fluent programming style of `DSLContext` is remarkable. This isn't your standard builder-pattern style chain of method calls. This is a true DSL that will prevent you from accidentally (or even purposefully) constructing incorrect SQL. Each method call in the chain is possible only if the previous method call will correctly support it. Truly, you'll need to work super hard to construct incorrect SQL in jOOQ. Because the DSLContext usage is mostly stateful, you'll need to be more conscious of thread safety when using this class.

3. `org.jooq.conf.Settings`

   The `Settings` class will let you further customize your jOOQ context with simple, straightforward parameters that change the behavior of the API. An instance of this class can help you control factors like

   - The kind of JDBC `Statement` that's jOOQ uses - a regular Statement or a `PreparedStatement`.[1]
   - Mapping different schemas in same jOOQ context.
   - Controlling the logging of SQL statements being executed by jOOQ.
   - Disabling support for Java Persistence API (JPA) annotations, for a slight performance improvement.
   - Configuring the behavior of jOOQ's internal SQL parser - for example, to set it up for a specific SQL dialect. This would apply if you're using jOOQ to process raw SQL strings instead of its typesafe options.
   - Configuring JDBC-specific options like `queryTimeout` and `maxRows`.
   - Configuring batch size for batch operations.

4. `org.jooq.Configuration`

   The `Configuration` class contains the core configuration items that control how your jOOQ runtime behaves. `Configuration` is responsible for managing your database connection, plugging

[1] Unlike with standard usage of `PreparedStatement` vs. `Statement`, you're at no greater risk of SQL injection by selecting one or the other in jOOQ. The difference here is largely performance related, where `PreparedStatement` queries are cacheable by the RDBMs.

into the jOOQ engine to customize its behavior on a broader scope than just individual SQL queries. `org.jooq.Configuration` provides methods that allow you to plug in custom code that can replace or support standard jOOQ functionality altogether.

5. `org.jooq.meta.jaxb.Configuration`

   You'll be using this class to control the code generation feature of jOOQ. It's a direct analog of a jOOQ configuration file named `library.xml`. Generating java representations of your database tables, rows, columns, stored procedures, etc. is a major feature of the jOOQ platform, and I'm looking forward to getting to that part of this book! Yes: `org.jooq.Configuration` vs. `org.jooq.meta.jaxb.Configuration` could lead to some awkwardness.

Which import statement is wearing it best? Trick question: it's Becky.

Eden Auto Mart

I'm going to use a car dealership as the background for all the examples that I'll be showing throughout this book. Eden Auto is a car dealer that sells new and used cars via a web portal so that customers can

- Search for cars by many different attributes of the vehicle itself
- Search across brands of vehicle
- Search across a price range

On the back end, the staff of Eden Auto can

- Upload cars for sale
- Update existing prices and other attributes
- Remove cars from the inventory
- View existing inventory
- Search the inventory for cars by many attributes

We're going to use a relatively simplified data model here just for the purpose of demonstrating specific jOOQ features. Here's what that looks like.

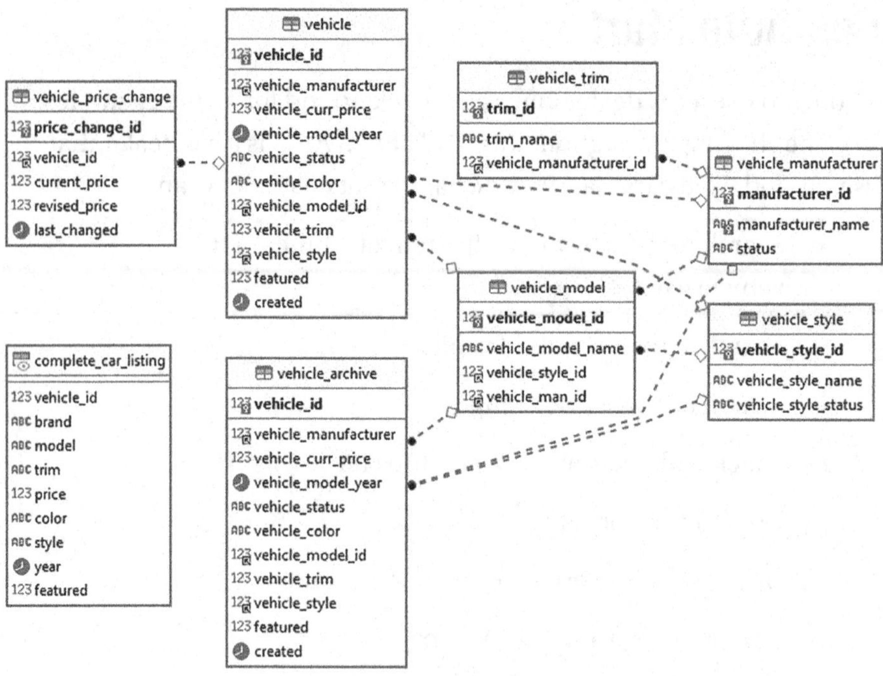

To help run some of the examples in this book, you can bootstrap with the database definition that comes with the code sample attached to this book.

Setting Up jOOQ

To start, you'll need to reckon with the particulars of the RDBMS you're going to be working with. A lot of the beauty of jOOQ is how it allows you to use pretty much any database with tons of convenience. Features that aren't available in your chosen database can be emulated transparently by the jOOQ API. For this book, I'll be doing most of the demonstrations with MySQL, with some detours into the some other popular databases every now and then.

I'll need MySQL's Connector/J database driver as a dependency:

```
<dependency>
        <groupId>mysql</groupId>
        <artifactId>mysql-connector-java</artifactId>
        <version>${mysql-driver-version}</version>
</dependency>
```

On to the actual setup of jOOQ, where things get a little...interesting. See, jOOQ is offered as both free and commercial software – a dual licensing model. The license model determines (among other things)

- The SQL dialects you can use in your application
- The way you set jOOQ up in your project

For the free version of jOOQ, it's a straightforward maven entry, because it's publicly available in the global Maven repo:

```
<dependency>
    <groupId>org.jooq</groupId>
    <artifactId>jooq</artifactId>
    <version>3.15.1</version>
</dependency>
<dependency>
    <groupId>org.jooq</groupId>
    <artifactId>jooq-meta</artifactId>
    <version>3.15.1</version>
</dependency>
<dependency>
    <groupId>org.jooq</groupId>
    <artifactId>jooq-codegen</artifactId>
    <version>3.15.1</version>
</dependency>
```

Simple enough, yes? Bueno. When you've paid for the commercial version however, things get a bit more manual. Here's how.

Install Dependencies for Commercial-Licensed jOOQ

The dependencies for the commercial[2] version of jOOQ aren't available in public repositories because, well, they're not free and thus not available for public download. To get the commercial version (or a trial), visit `www.jooq.org/download/versions` to download the version that matches your version of Java.

`<screenshot of versions page>`

Once you've plugged in your particulars, you'll be prompted to download a zip file containing

- Sources
- Compiled JARs
- Helper scripts

There are two helper scripts in the zip file:

- `maven-deploy`
- `maven-install`

Both helpers do the same thing with different targets: build and install the jOOQ jars into repositories. `maven-deploy` will set up the jOOQ JARs in a remote repository, so reach for that to set up the JAR in a central

[2] Commercial = Express, Professional, and Enterprise licenses

Artifactory or similar dependency repository. For a local maven repository, reach for the `maven-install` script and you're sorted. After all this scriptin' action, you can then add commercial jOOQ to your project like so:

```
<dependency>
   <groupId>org.jooq.pro</groupId>
   <artifactId>jooq</artifactId>
   <version>3.15.1</version>
</dependency>
<dependency>
   <groupId>org.jooq.pro</groupId>
   <artifactId>jooq-meta</artifactId>
   <version>3.15.1</version>
</dependency>
<dependency>
   <groupId>org.jooq.pro</groupId>
   <artifactId>jooq-codegen</artifactId>
   <version>3.15.1</version>
</dependency>
```

It's the same `artifactId` as the free version, but with a different `groupId:org.jooq.pro`. You can use a trial license for the commercial version of jOOQ by using `org.jooq.trial` for the `groupId`. The open source version is just as functional for many use cases, but with limited database vendor support and fewer features.[3] Also noteworthy that jOOQ's JDK support starts from JDK **6**, all the way up to the latest (17, as at the time of this writing). The older JDK versions are supported only with the paid version.

[3] www.jooq.org/download/#feature-comparison

Tip The trial version of the commercial jOOQ distribution will print a message indicating that yes, you are indeed on a trial version of the commercial jOOQ distribution. But rejoice, oh ye trial version users, for jOOQ is effective as it is generous: set the `org.jooq.no-logo=true` JVM property to disable the trial license warning message.

And that's it! jOOQ is set up in your project. Now to do stuff with it...

CRUD with jOOQ

With your jOOQ JARs in place, database driver configured, we now should start using this bad boy. We're going to need to acquire connections, load the driver, etc. If you haven't done database work in Java before now, I'm going to show you how the least cool way to do it in Java:

```
try(Connection connection = DriverManager.
getConnection("jdbc:mysql://localhost/test?user=testuser&
password=thisisaterriblepassword")){
      //more to come
}catch(SQLException sqlex){

}
```

The goal of the preceding snippet is to acquire a database connection; jOOQ can take care of everything following that. jOOQ on its own isn't too particular about how you acquire the connection if you follow some established commonsense principles. This isn't a particularly great way to do connection acquisition in a modern application. You should be using the `javax.sql.Datasource` class and connection pools instead of manually wrangling connection drivers. But more on that later. For now, we have a database connection, and thus it's time to start jOOQin.'

Remember `DSL` and `DSLContext` are the primary entry points into the jOOQ ecosystem.

Fundamentally, most jOOQ operations will begin with some variation of the following:

```
DSLContext context = DSL.using(connection,SQLDialect.MYSQL);
```

In the preceding sample

- I supply the JDBC connection object (how I obtained the connection isn't important for now).
- And I supply a dialect from the `SQLDialect` enum to pass to the `DSLContext`.

According to the manufacturer, `DSLContext` isn't always guaranteed to be thread safe – it is a context object after all. However, if you personally take sensible precautions, you could enjoy thread safety with this class. Having provided `DSLContext` with a valid JDBC connection, you can now get into the nitty-gritty of **Create, Read, Update, and Delete** (CRUD). But first, a word from our sponsors...

Your SQL Dialect and You

While it's not mandatory to set a SQL dialect on your `DSLContext`, it's ideal you do. See, RDBMSes have many different quirks as I've already covered. Some of them are glaring and easily detectable. Other more trivial ones can trip you up unexpectedly. Take my choice of MySQL database, for example:

- Identifier Style: You might already be aware that different databases use different quote styles for identifiers. Because of this, standard SQL will have

```
select "e"."first_name", "e"."last_name" from
"employees" "e"
```

but MySQL has[4]

```
select `e`.`first_name`, `e`.`last_name` from
`employees` `e`
```

Backticks, instead of double quotes. This isn't to say that you'll be denied a lot of functionality without specifying a dialect. *Au contraire,* jOOQ will routinely go out of its way to emulate features that are supported natively by one or two databases but not by others, for example:

- Returning Keys: A SQL `INSERT... RETURNING` in PostgreSQL will return with the generated keys of the rows that have been inserted, and it's immediately available because of the insert operation. This is provided for by the SQL standard but isn't uniformly supported by databases. Among those that do, the implementation details vary. jOOQ can emulate this feature for you, regardless of the kind of database you're using. Fair warning here: depending on the kind of support your database has for it, jOOQ may be forced to silently initiate a second `SELECT` trip to the database to retrieve the generated keys.

[4] You can control this behavior by configuring the ANSI_QUOTES option for MySQL (`https://dev.mysql.com/doc/refman/8.0/en/sql-mode.html#sqlmode_ansi_quotes`).

- Dummy Tables: Different SQL databases allow you to select from dummy tables in different ways. Oracle has the famous `DUAL` table, for example. Sometimes, you just want to run some functions, but then your RDBMS requires you to specify a `FROM` clause – without offering a dummy table. jOOQ supports select statements without from clauses so you can do whatever sorcery you like, with or without a dummy table.

If you don't specify any dialect, you'll get the default `SQLDialect.DEFAULT` and then *via con Dios*![5]

Caution For your own peace of mind, go ahead and configure the `org.jooq.conf.Settings.backslashEscaping` property on your `Settings` object. MySQL and some versions of PostgreSQL support non-standard escape characters that can cause you a lot of grief when you least expect it. This property lets jOOQ properly handle this "feature" from MySQL.

---

[5] Lukas: If you provide `DSL.using(connection)`, then jOOQ will try to "guess" the appropriate SQLDialect from the JDBC DatabaseMetaData – `https://docs.oracle.com/javase/7/docs/api/java/sql/DatabaseMetaData.html`

And that has been the word from our sponsors!

Thank you for reading!

Okay, let's now get into...

Tools of CRUD in jOOQ

jOOQ offers *a lot* of fine-grained control of the entirety of a SQL statement - really, an insane amount of control at your fingertips. Starting from a pretty high level, I'm going to go over some of the key components that you're going to need to get familiar with if you're going to CRUD with jOOQ:

1. `org.jooq.Query`

   This is the fundamental unit of a SQL query in jOOQ-land. It extends `org.jooq.Statement` along with the jOOQ representations of other executable database units of work like stored procedures and functions. You can use it to execute

- Plaintext SQL that you pass in as an argument
- Strongly typed SQL fragments constructed using the jOOQ API

It can help you convert a manually derived plaintext SQL to a jOOQ DSL-typed SQL object and vice versa. It's the quintessential representation of a SQL query; use it to execute SQL statements when you don't need a return value from the operation (like Data Manipulation Language operations).

2. `org.jooq.ResultQuery`

   This class is basically `org.jooq.Query`, but you can obtain return values from it, like query results. Some notable qualities of this class:

   - It's efficient in that it doesn't need an active database connection to provide access to its results - once the query has been successfully executed, the connection can be discharged, and you can interrogate your instance of this class for all the query results.[6] This is useful when you are resource conscious and want to protect against long-running transactions. Compare this feature to detached hibernate entities.

[6] Lukas: It's time to deprecate that ancient feature of the mutable `ResultQuery`! This won't be available in the future anymore. In the future, only `ResultQuery::fetch` and similar methods should be used.

- It provides typed or untyped result fetching, comparable to what you get with the Spring `JdbcTemplate`.
- It maps to custom Plain Old Java Objects (POJOs).
- It supports lazy fetching of query results using an underlying database cursor. If you go this route, be mindful that it depends on maintaining the live database connection.
- It supports reactive programming.
- It supports streams via the `java.util.stream.Stream` interface.

All told, it provides a very flexible interface for wrangling query results. It's as powerful as anything you'll get with Hibernate or JPA, with much less verbosity than what JDBC would put you through for the same outcomes.

3. `org.jooq.QueryPart`

   On its own, not too powerful. But it's the parent type of a large suite of classes that help you construct complex queries dynamically. Compare this class to the `CriteriaBuilder` in JPA or the `Specification` from Spring Data JPA. To be clear, this API is exorbitantly richer than either of what JPA or Spring Data provide. Pretty much every fragment of the SQL standard specification can be represented as variant of `org.jooq.QueryPart`, for example:

   - `Field` to represent a field that can be used anywhere, in any kind of SQL statement
   - `Condition` to represent predicates like `WHERE`, `HAVING`

- `SQL` to represent a whole SQL statement or fragment
- `Table` to represent a whole table, wherever a table is needed in your query
- `Row` to represent tuples (or in layman's terms, something like a row of a table)
- `Field` to represent…yes, a column
- `Constraint` to represent a database constraint, as part of a DDL operation
- `Schema` to represent the schema part of any SQL query

This is a small sampling of the level of granularity that `QueryPart` can get to. Heck, there are even more specialized versions of each of these that offer specific advantages. If you've ever dreamed of being able to support a powerful dynamic filter as part of a search function, but without concatenating strings of ugly plaintext SQL all over the place, this is where you want to be.

You get powerful and reusable components to construct any kind of SQL use case you can dream of. When you use jOOQ's code generation feature, the artifacts that are generated inherit from this family as well. Incredibly powerful stuff.

4. `org.jooq.Record`

   This is the parent class for representations of a row of data from a table. This is in addition to jOOQ's capability to work with your custom POJOs and Data Transfer Object (DTOs). You get the following specializations of this class, among others:

- `org.jooq.UpdatableRecord`

  Compare this to a hibernate entity: it's a live object that remains bound to the underlying database row for its lifetime. This way, you can modify the refresh, modify, or delete an instance of `UpdatableRecord`, and it can reflect the change in the underlying table.

  When you use jOOQ's generated artifacts, you'll get classes that extend this one by default. This applies only to tables with primary keys – without primary keys, you'll get a different generated artifact. Additionally, you'll be able to navigate to related entities of an updatable record using its foreign key references. Just like JPA! But better! `UpdatableRecords` isn't cached, so you can be guaranteed you're always working with up-to-date, live data. It's also easier on RAM.

- `org.jooq.TableRecord`

  This is what you get instead of an `UpdatableRecord` when the underlying table doesn't use a primary key. So, you won't be able to refresh or update parts of these records; you'll still be able to insert rows with this class, however.

- `org.jooq.RecordMapper`

  Use this class for finer-grained control over the translation of your query results. Compare this to Spring's `RowMapper` class.

5. `org.jooq.Result`

   This is a container class to hold query results. In practice, your `Record` objects or lists of `Record` objects will be wrapped by an instance of `Result`. It's a `Collection`, `Iterable`, `List`, and `Serializable` – so you can do a lot with it (functional, reactive, and JDK Stream programming). It's interoperable with JDBC's `ResultSet` as well. It has the added advantage that it doesn't hang on to the open database connection like the `ResultSet`.

   Do you feel the power?

Do you?

I hope so because we're about to use this power for...

Select Statements

Let's say hello to our old friend, DSLContext:

```
DSLContext context = DSL.using(connection,SQLDialect.MYSQL);
//hullo!
```

For starters, we need to be able to run a vanilla SQL query. Here's what that looks like. Given a custom CompleteVehicleRecord POJO that I manually created:

```
public class CompleteVehicleRecord {
    Long id;
    String brand;
    String model;
    String trim;
    BigDecimal price;
    String color;
//getters, setters, toString and hashCode;
}
```

This class encapsulates an individual vehicle for sale in the database. I'm interested in selecting all available vehicles in the database. In SQL terms, it would look something like this:

```
SELECT * FROM edens_car.complete_car_listing;
```

Here's how it looks in terms of jOOQ:

```
DSLContext context = DSL.using(connection,SQLDialect.MYSQL);
                ResultQuery resultQuery = context.
                resultQuery("SELECT * FROM edens_car.complete_
                car_listing"); (1)
                List<CompleteVehicleRecord> allVehicles =
                resultQuery.fetchInto(CompleteVehicleRecord.
                class); (2)
                logger.info(allVehicles.toString());
```

Nothing fancy, just your vanilla select statement.[7] Here:

1. I pass in my plaintext SQL to the `DSLContext#resultQuery` method to retrieve all available cars. I choose to obtain a `ResultQuery` from the execution because I want to get the results from it. If this were a query that I didn't want results from (say, a delete or insert statement), I would have just used the `Query` class instead.

2. On the `ResultQuery`, I call the `fetchInto` method to automatically map each row from the results by

    - Mapping each row returned into an instance of `CompleteVehicleRecord`

    - Putting all the mapped rows into a list of `CompleteVehicleRecord`

    jOOQ is able to infer the mappings between the fields of my POJO and the columns returned in the jOOQ select statement. As you've probably guessed,

[7] This isn't the best type of select statement - you want to be specific about the columns you want in a query. However, the focus of this book is on working the best way with jOOQ, not necessarily SQL.

the names of the class variables and the columns in the SQL query need to match. There are other ways to deliberately establish this mapping; we'll check them out later in the book.

Alternatively, I could use the fetchMany method so I can take more control of mapping each row:

```
resultQuery.fetchMany().forEach(results -> {
                    //container for all the results
                    results.forEach(record -> {
          logger.info("New result row");
          logger.info("Brand: {}",record.getValue("brand"));
          logger.info("Model: {}",record.getValue("model"));
          logger.info("Trim: {}",record.getValue("trim"));
          logger.info("Color: {}",record.getValue("color"));
          logger.info("Price: {}",record.getValue("price"));
          });
     });
```

The fetchMany method is guaranteed to never return a null, so don't base any logic around null checking. If I were expecting just one result, I'd have used the fetch method instead. You can also use this method to execute multiple SQL statements in the same execution - this is an understated but very powerful method. Keep reading for more uses of this method.

I could also supply query parameters to my SQL statement like so:

```
ResultQuery resultQuery2 = context.resultQuery("SELECT * FROM
edens_car.complete_car_listing where color = ?","BLUE");
```

Query bindings – the bit where I say "BLUE" – are how you provide parameters to your SQL statements. This approach to binding query parameters offers some SQL injection protection when combined with the PreparedStatement configuration on the underlying DSLContext API.

Remember jOOQ can use the JDBC `PreparedStatement` component by default if you configure it to, so your parameters can still benefit from the inbuilt SQL injection safety in that component. You can also cache the underlying `PreparedStatement` by calling the `keepStatement` method on the `Query` object; this gives a performance boost for frequently used, stable queries.

Let's look at some much cooler, purpose-driven uses of the jOOQ library.

The SELECT DSL

jOOQ offers to protect you from malformed SQL, SQL injection, missing parameters, and overly restrictive and ugly code. It does this in incrementally powerful ways, so you can move at your own pace. Let's revisit the `DSLContext` in the *context* (hehe) of the `SELECT` statement again:

```
SELECT * FROM edens_car.complete_car_listing
```

I can rewrite the preceding plaintext `SELECT *` like so:

```
DSLContext context = DSL.using(connection,SQLDialect.MYSQL);
List<CompleteVehicleRecord> allVehicles =
context.select().from(table("complete_car_listing")).
fetchInto(CompleteVehicleRecord.class);
```

Both statements are functionally identical and will return the same results:

1. `select()` is jOOQ shorthand for `SELECT *`. jOOQ is chock full of shorthand like this that helps to cut down on verbosity.

2. `table` is a function from the `DSL` class, imported statically. It helps me convert the plaintext SQL name of the table to an instance of `org.jooq.Table`. The effect of this is that jOOQ can treat my statement almost like a typesafe version.

What I have here is still partially in the plain SQL realm, and as a result, I must use helper functions in the `DSL` class to sort of translate my plain SQL into something like the typesafe API of jOOQ. Think about it this way: if I'm still using plain strings in key parts of my select statement, jOOQ still can't 100% guarantee correctness in many aspects. This DSL makes sure that my queries are syntactically correct, but it can't make guarantees as to the semantic correctness – I'm still allowed to make typographical errors in the names of tables, columns, etc.

Now, the `SELECT *` is a tad wasteful, so I would typically prefer to use SQL projections to be more succinct about which columns I'm interested in:

```
List<CompleteVehicleRecord> allVehicles = context.select(field(
name("brand")),field("model"),field("price"))
            .from(table("complete_car_listing"))
            .fetchInto(CompleteVehicleRecord.class);
```

Like the `table` function, `field` comes from `org.jooq.impl.DSL`, and I'm using it here just to bridge the gap between my hand-crafted, gluten-free, artisanal SQL and jOOQ's strongly typed, DSL-centric world. The `field` function converts my raw SQL field name into an instance of `org.jooq.Field`, `table` will convert my table name to an `org.jooq.Table`, and `name` will create an instance of `org.jooq.Name`. All of them are `QueryParts`, and they form the basis of being able to construct powerful and complex queries.

Another way I can construct a select statement is by setting up the relevant fields individually:

```
Field<?> brandField = field("brand"); (1)
Field<?> modelField = field("model",String.class);
Field<BigDecimal> priceField = field("price", BigDecimal.
class); (2)
List<Field> fieldList = Arrays.asList(brandField,modelField,
priceField); (3)

List<CompleteVehicleRecord> allVehicles = context.select(fieldList)
                        .from(table("complete_car_listing"))
                        .fetchInto(CompleteVehicleRecord.class);
logger.info(allVehicles.toString());
```

In the preceding snippet

1. I manually construct an instance of `org.jooq.Field` from plaintext "SQL." Note the wildcard syntax I'm using here, `Field<?>`. It's purposeful: jOOQ is all about type safety, so at many turns it would like *something* resembling type information. Get used to specifying the types of `Field`, and when you can't provide it, use the wildcard.

2. Even better, I can supply type information on both sides of the variable declaration. `Field` can be typesafe, and the `field` function can be provided with type safety info. Because I'm not quite using jOOQ's typesafe-generated code, any additional information I can provide along with my hand-carved SQL can be used by jOOQ to protect the integrity and reliability of my SQL statements.

3. I can then stash all the necessary fields into a standard java `List`.

4. The list can then be used by any part of a jOOQ query.

It's really powerful stuff, this mechanism. I can gain a lot of reusability and flexibility in my code with this approach, even when I'm not generating code with jOOQ. I'll be showing you more scenarios of this sort of reusability as we go – this is the power of the `QueryParts` components in the jOOQ library.

Tip Use the `DSL#name` function to handle your raw SQL identifiers in a schema-safe format. It can also provide SQL injection protection when quoting is enabled for the `DSLContext`.

Other options with `select` include

- Running a `SELECT DISTINCT` with `selectDistinct`.
- Selecting from dummy tables (like DUAL for Oracle, or nothing for PostgreSQL) by running `selectOne().fetch`. This feature depends on a correctly set SQLDialect parameter in the `DSLContext`.
- Combining individually constructed `SELECT` statements; more on this later in the book.

What if I want a query within a query? You know it! It's time to dig into `SELECT`ing...

With the WHERE DSL

Having decided on which columns I'm interested in from my vehicle table, I want to be more restrictive about which rows come back – enter the almighty WHERE clause. Here's what could look like:

```
List<CompleteVehicleRecord> allVehicles = context.select(
field("brand"),field("model"),field("price"))
              .from(table("complete_car_listing"))
              .where(condition("color = 'BLUE'"))
              .fetchInto(CompleteVehicleRecord.class);
```

I'm technically using the DSL here, though not to great effect. This is still very much a plaintext SQL where clause, and I should be ashamed of myself – which I am. So, I'll try again with bind variables like so:

```
context.select(field(name("brand")),field("model"),field("price"))
             .from(table("complete_car_listing"))
             .where(condition("color = ?","BLUE"))
             .fetchInto(CompleteVehicleRecord.class);
```

This is a bit safer from a SQL injection perspective. I'm still responsible for making sure that the syntax of the preceding snippet is correct and will come out properly when the whole thing's stitched together. Overall, this still reads better and more fluently. One of the things I love about the jOOQ DSL is that it's devoted to making sure one doesn't make mistakes in constructing SQL. For example, it'll be impossible for me to use the where node anywhere but after a from clause. So even if I choose to keep using plain SQL at specific intervals, I still get some protection from the fact that my SQL is syntactically correct. Additionally, I'm benefitting from the resource-efficient approach that jOOQ uses with handling JDBC connections.

Pop Quiz: What's the difference between the `org.jooq.impl.Settings` and the `org.jooq.impl.Configuration` classes?

Answer: The `Settings` class will allow you to change predefined behavior of the jOOQ runtime by simply changing a setting. The `Configuration` class provides access to extension points in jOOQ, for you to supply custom code that will replace or alter the behavior of the jOOQ runtime.

Now, for one of my favorite components of the jOOQ API...

Conditions

The arguments you supply to the `from` node of the `select` DSL are in reality instances of `org.jooq.Condition`. The `Condition` is a powerful class that lets you compose simple or complex predicate components. You can then attach to almost anywhere in the SQL structure that accepts conditional logic. So, I can write a `Condition` like this:

```
Condition colourCondition = condition("color = ?","BLUE");
...and then pass that into my constructed {select} execution:
context.select(fieldList)
          .from(table("complete_car_listing"))
          .where(colourCondition)
          .fetchInto(CompleteVehicleRecord.class);
```

Being able to dynamically construct parts of a whole SQL query is fundamental to the way jOOQ works.[8] You'll see different flavors of this statement as you go through this book, but every section of your SQL statement can be constructed independently of the rest. Your SELECT,

[8] www.jooq.org/doc/latest/manual/sql-building/dynamic-sql

WHERE, or HAVING clauses, etc? You can build them independently and later stitch them together.

Things can get even *more* flexible with some of the convenience utilities that jOOQ offers with Condition. For example, I can construct a jOOQ query with an optional WHERE clause like so. Consider the following jOOQ query that I use to select car details from the complete_car_listing table:

```
List<CompleteVehicleRecord> allVehicles = context.select(field(
name("brand")),field("model"),field("price"))
            .from(table("complete_car_listing"))
            .where(colourCondition)
            .fetchInto(CompleteVehicleRecord.class);
```

So, this looks like the same jOOQ query that you've been seeing so far, yes? How can I make this query work with an *optional* WHERE clause? This way, I can reuse the same query whether the website user selects filter criteria or not. So, consider a hypothetical user interface like this:

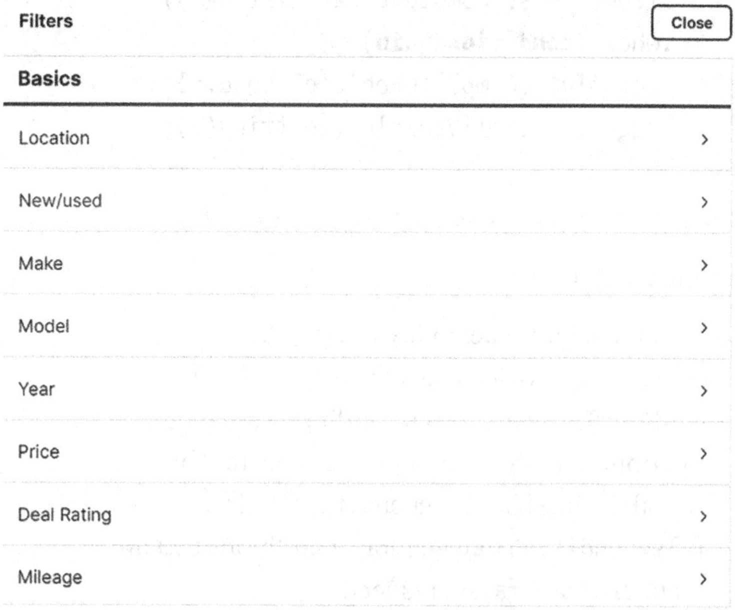

Filters	Close
Basics	
Location	›
New/used	›
Make	›
Model	›
Year	›
Price	›
Deal Rating	›
Mileage	›

The filter criteria selection box for the ***Eden*** *Auto website*

There are multiple options to filter search results by. You need to be able to use the same jOOQ statement, whether the user selects any of the filter criteria or not. Here's what that jOOQ query looks like:

```
public static void selectWithOptionalCondition(boolean
hasFilter, Map<?,Object> filterValues) throws SQLException{
        ...
        Condition conditionChain = DSL.noCondition(); (1)
        if (hasFilter) {
                for(String key: filterValues.keySet()){
                    conditionChainStub = conditionChainStub.
                    and(field(key).eq(filterValues.get(key))); (2)
                }
            }

     List<CompleteVehicleRecord> allVehicles = context.select
     (field(name("brand")),field("model"),field("price"))
           .from(table("complete_car_listing"))
           .where(conditionChain)
           .fetchInto(CompleteVehicleRecord.class);
           logger.info(allVehicles.toString());
           }
}
```

Here's the breakdown:

1. To get maximum flexibility with the `Condition}` class, jOOQ provides the `DSL#noCondition()` method. This generates a condition stub that I can optionally chain other conditions to. There are others like `DSL.trueCondition` and `DSL.falseCondition` that generate conditions that are set to `true` and `false`, respectively.

2. `Condition` provides the and operator. Using this facility, I can combine the filter conditions that are passed in from the web tier, if they exist. If no filter parameters are passed in (i.e., `hasFilter` is false), no `WHERE` condition will be applied to the eventual SQL statement that's executed. Otherwise, the constructed `Conditions` will be applied.

As you can probably tell, the `Condition` class provides all the boolean operators you'd need: `or`, `not`, `exists`, as well as all the useful permutations of all of them. Let's not forget the comparison operators on the `Field` class:

```
Condition condition = field(name("price")).
greaterOrEqual(BigDecimal.valueOf(360000));
```

That's right: the `Field` class itself can yield conditions by virtue of the many, many comparison operators available on the class itself.

For *even* more flexibility in constructing your conditions, check out the `CustomCondition` class:

```
CustomCondition customCondition = CustomCondition.of(
conditionChain ->{
               conditionChain.sql("color ='BLUE'")
                      .sql(" AND price < 35000");
          });
```

`CustomCondition` provides the opportunity to perform complex logic in the process of building the condition. By providing a functional interface that accepts a lambda, you can take even more control of the process. It still yields an object that you can combine with any other flavor of condition you have.

Flexibility!

Pro Tip The `Field` class offers the `isNull` and `isNotNull` for all your null comparison needs. Stay safe; use the appropriate null comparison methods. Note that these are offered in addition to the database functions dedicated to handling nulls (e.g., `COALESCE`, `NVL`, etc.).

Subqueries

As an example, I want to search for vehicles of a specific manufacturer that have had recent price reductions – because I'm thinking price reductions means that nobody's buying the car, and the dealer might be willing to give me a bargain.[9] Here's what the plain SQL query would look like:

```
SELECT *                                (1)
FROM  complete_car_listing ccl
WHERE (ccl.vehicle_id , ccl.price) IN  (1a)
          (SELECT vpc.vehicle_id, vpc.revised_price     (2)
              FROM vehicle_price_change vpc
              WHERE vpc.revised_price < vpc.current_price
              AND (vpc.vehicle_id , vpc.last_changed) IN
          (SELECT vc.vehicle_id, MAX(vc.last_changed) (2a)
              FROM vehicle_price_change vc
              GROUP BY vc.vehicle_id))
```

[9] Editor's note: smh.

This query (technically one main query and two subqueries) will

1. Retrieve all the details of cars that are in the inventory

   a) I match the rows of the top-level query by using the SQL row value expression mechanism to compare against the results of the subquery.

2. Find cars that have a price that has been revised downward

   a) And of that subset, make sure that the most recent price revision was a reduction.

Caution "jOOQ is all about type safety" – me, a couple of pages ago. This is still true, but you'll see me skip some type safety convention in subsequent code samples, for example, using `field("price")` instead of `field("price",BigDecimal.class)`. This is largely to cut out a bunch of boilerplate code; the fundamental truth of type safety remains intact. **Using plaintext SQL is still an inferior option compared to generating typesafe code with jOOQ**. As you continue your jOOQ journey, plaintext SQL will start yielding issues and weirdness as a direct consequence of not having type safety. Please bear this in mind as you proceed.

To be clear, there are other, probably better ways of achieving the same results: joins, window functions (more on those later), and others. This is a contrived example to demonstrate some specifics of subqueries in jOOQ. If you run this query against the schema that accompanies this book, you should get results that look something like the following:

```
'3', 'Lexus', 'ES 350', 'BASE', '36000.0000', 'BLUE', 'Car', 2018
'4', 'Acura', 'MDX', 'SPORT', '50000.0000', 'BLUE', 'Car', 2018
```

What would this look like in jOOQ? To begin, I'm going to declare a couple of fields and tables for easier reuse in the queries I'm going to be constructing:

```
Field<Long> vehicleId = field(name("vehicle_id"),Long.class);
Field<BigDecimal> vehicleRevisedPrice = field(name("revised_
price"),BigDecimal.class);
Field<BigDecimal> vehicleCurrentPrice = field(name("current_
price"),BigDecimal.class);
Field<BigDecimal> price = field(name("price"),BigDecimal.class);
Table vehiclePriceChange = table(name("vehicle_price_change"));
Field<LocalDateTime> lastPriceUpdate = field(name("last_
changed"),LocalDateTime.class);
```

These being set up, I can go about setting up the actual queries:

```
final SelectCorrelatedSubqueryStep<Record2<Long,
LocalDateTime>> mostRecentPriceChange = context.
select(vehicleId, max(lastPriceUpdate)).
from(vehiclePriceChange).groupBy(vehicleId); (1)
final SelectConditionStep<Record2<Long, BigDecimal>>
mostRecentPriceReduction =  context.select(vehicleId,vehicle
RevisedPrice)
.from(vehiclePriceChange)
.where(vehicleRevisedPrice.lessThan(vehicleCurrentPrice))
.and(row(vehicleId, lastPriceUpdate).
in(mostRecentPriceChange)); (2)
```

Here's what I'm up to with these queries:

1. I construct the query that will provide the most recent price change per vehicle ID. The max method comes from the trusty DSL class.

2. Next, I use the query from (1) to construct the query that gets the vehicles that have had only price reductions recently. The `row` method is also from the `DSL` class to enable tuple comparisons against the subquery.

What's happened is that I've separately constructed jOOQ SQL queries to facilitate reuse. Now, a word from our sponsors: "jOOQ doesn't officially recommend this approach for reusing queries, in part because of mutability concerns." If the readability wouldn't be too poor, you're safer inlining the subquery into the main query. For the purposes of this demonstration, I've broken the subqueries out and made them `final`.

Right, subqueries constructed, let's crack on with using them:

```
SelectConditionStep<Record> potentialDealsQuery = context.
select().from(table(name("complete_car_listing")))
                              .where(row(vehicleId,price)
                              .in(mostRecentPriceReduction));
String sql = potentialDealsQuery.getSQL();
logger.info(sql);
```

The `potentialDealsQuery` uses the `mostRecentPriceReduction` subquery to get the cars that have had recent price reductions. To see what the generated query looks like, I can get the SQL off the `potentialDealsQuery` with the `getSQL` method. Here's the result:

```
select * from `complete_car_listing` where (`vehicle_id`,
`price`) in (select `vehicle_id`, `revised_price` from
`vehicle_price_change` where (`revised_price` < `current_price`
and (`vehicle_id`, `last_changed`) in (select `vehicle_id`,
max(`last_changed`) from `vehicle_price_change` group by
`vehicle_id`)))
```

Largely a faithful translation of the raw SQL intent, yeah? As usual, I'll fetch the results:

```
List<CompleteVehicleRecord> potentialDeals =
potentialDealsQuery.fetchInto(CompleteVehicleRecord.class);
```

That was a lot huh? You've bought the book; please feel free to go over this section again if it didn't land the first go around. I've introduced multiple concepts here at once that will be helpful overall in jOOQ. It's completely understandable if it doesn't all click at first read through 😊

Pop Quiz: How would you safely refer to fields when working with your plaintext SQL in jOOQ?

a) `Field myField = field("myField")`

b) `Field <?> myField = field("myField")`

c) `Field<Integer> myField = field("myField", Integer.class)`

Answer: (b) and (c) are the recommended approaches; (c) more preferably!

Conditional Logic in Queries

If you'd like to get fancy, you could have some sophisticated conditional logic in your SQL. In case I didn't make it clear before now: I'm firmly in the camp of "Let the database do the things the database is very good at, with maintainability in mind." To that end, my soul frowns when I see code that

- First retrieves query results into the application layer
- Then performs complex analysis that the database is otherwise exceedingly good at

So, we've established I'm a cheapskate and I'm always looking for a good deal on a car – these two are mutually exclusive. We've seen how to find cars with price reductions, and therefore might probably offer good deals. How good of a deal are we talking about here? I'd say

- 5% reduction, "Okay"
- 10% reduction, "Good"
- Above 10%? "Great!"

How would a SQL query present this? With the CASE function:

```
SELECT vpc.vehicle_id 'vehicle', vpc.current_price 'old price',
vpc.revised_price 'new price', max(last_changed) 'last price
update',
case when ((vpc.current_price - vpc.revised_price)/vpc.current_
price)*100 <=5 then 'OK'
         when ((vpc.current_price - vpc.revised_price)/
         vpc.current_price)*100 BETWEEN 5 AND 10 then 'GOOD'
         when ((vpc.current_price - vpc.revised_price)/
         vpc.current_price)*100 > 10 then 'GREAT'
```

else 'NO DEAL'

```
 end as 'deal'
FROM vehicle_price_change vpc
WHERE vpc.revised_price < vpc.current_price
GROUP BY vpc.vehicle_id, vpc.current_price, vpc.revised_price
```

For results that look like this:

```
# vehicle   old price    new price   deal
2           48000.0000   47380.00    OK
3           37565.0000   36000.00    OK
4           55342.0000   50000.00    GOOD
```

You probably know where this is going: how to represent this in jOOQ? Hang on to your keyboard:

```
context.select(vehicleId, vehicleCurrentPrice,
vehicleRevisedPrice, max(field("last_changed")),
                    when((vehicleCurrentPrice.subtract(vehicle
                    RevisedPrice))
                            .divide(vehicleCurrentPrice)
                            .multiply(100)
                            .lessOrEqual(BigDecimal.
                            valueOf(5)), "OK") (1)
                            .when(condition("((current_price -
                            revised_price)/current_price)*100
                            BETWEEN 5 AND 10"), "GOOD") (2)
                            .when(condition("((current_price -
                            revised_price)/current_price)*100 >
                            10"), "GREAT")
                            .otherwise("NO DEAL") (3)
                            .as("deal")) (4)
   .from(table("vehicle_price_change"))
   .where(vehicleRevisedPrice.lessThan(vehicleCurrentPrice))
   .groupBy(vehicleId)
         .fetch()
             .forEach(result -> {
                   logger.info("Vehicle Id: {} | Revised
                   Price: {} | Former Price: {}| Deal
                   Rating: {}", result.get(vehicleId),
                   result.get(vehicleRevisedPrice), result.
                   get(vehicleCurrentPrice), result.get("deal"));
                    });
       }
   }
```

Let's break this down:

1. Skipping past the other fields in the SELECT statement: I start with the when method, into which I use various methods of the Field class to construct the arithmetic that constitutes a deal. I do the arithmetic and then pass "OK" as the outcome of this when condition. Note how this doesn't actually feature the case keyword/method.

   - Alternatively, if you're looking to use the form of CASE (column), you'd begin with the choose method instead of when.

2. I use the plain SQL option for this when to demonstrate the utter flexibility that's available. You'll notice that I'm using the condition method here, because that's essentially what the when method needs: a jOOQ Condition. This means that you can construct and reuse Conditions before you need them in a select statement.

3. The otherwise method gives me the ELSE clause for my CASE – the catch-all.

4. as gives me an alias for the whole case statement.

All of this gives me the following output:

```
Vehicle Id: 2 | Revised Price: 47380.0000 | Former Price:
48000.0000| Deal Rating: OK
Vehicle Id: 3 | Revised Price: 36000.0000 | Former Price:
37565.0000| Deal Rating: OK
Vehicle Id: 4 | Revised Price: 50000.0000 | Former Price:
55342.0000| Deal Rating: GOOD
```

This is one of my favorite demonstrations in this book, because it shows just how flexible jOOQ gets.

Everything fits everywhere, and you can compose SQL statements from any level of granularity; and this isn't even a complicated example of that power.

jOOQ supports some of the more vendor-specific conditional functions like

- DECODE
- COALESCE
- NVL
- NVL2
- IIF and IF
- NULLIF

All of these are available as functions in...you guessed it: the DSL class!

With the… WITH Clause

If I were interested in calculating the median price of all the vehicles in the database, I would have to get a little creative. See, it's not a standard SQL function (yet). PostgreSQL supports it somewhat natively,[10] but for most other databases, it's going to take some tinkering. In my experience, "tinkering" in SQL tends to require fairly unsightly SQL; SQL that I'd really want to be able to reuse elsewhere in my SQL query. Here's what it looks like when I use SQL window functions (more on those later) to calculate the median price of all the vehicles in the inventory:

```
WITH median_cte as(SELECT ROUND(AVG(price)) median
                   FROM (select price, ROW_NUMBER() OVER (ORDER
                   BY price ASC) AS rowpos, COUNT(*) OVER () AS
                   total_cars from complete_car_listing) price_mod
```

[10] www.postgresql.org/docs/current/functions-aggregate.html#FUNCTIONS-ORDEREDSET-TABLE

```
                    WHERE rowpos BETWEEN total_cars / 2.0 AND
                    total_cars / 2.0 + 1)
select brand, model, trim, price, CONCAT((ROUND((price -
median_cte.median)/price,2) * 100),'%') 'relative to median'
from complete_car_listing, median_cte
```

In MySQL, the WITH clause runs the median query once, stashing the result in a temporary "table."[11] I can then refer to the result in the subsequent SELECT statement using the name I specified, {median_cte}, almost like a table. This is what's known "in the biz" as a Common Table Expression (CTE) – which you may already know. What does that look like in jOOQ?

```
Field<BigDecimal> price = field("price",BigDecimal.class);
//define field and table for reuse
Table completeCarListing = table("complete_car_listing");
CommonTableExpression<Record1<BigDecimal>> medianCte =
name("median_cte") (1)
     .as(context.select(round(avg(price)).as("median")) (2)
                    .from(select(price, rowNumber().
                    over(orderBy(price.asc())).
                    as(name("rowpos")),
                                   count().over().as("total_
                                   cars"))
                                   .from(completeCarListing))
                                   .where("rowpos BETWEEN
                                   (total_cars / 2.0) AND
                                   (total_cars / 2.0 + 1)")
          );
```

[11] The mechanism for the WITH clause varies from database to database, but the outcomes are identical.

What voodoo is happening here? I'll tell you:

1. I construct an instance of `CommonTableExpression`, a variant of `org.jooq.Table`. This means that I can treat this object a lot like a standard SQL table in many instances. The logic in this block is a window function to calculate the median price of vehicles in the database. I cover window functions in a later section of this book; you can disregard it for now. What's most important is that I've encapsulated the query in here and named it `median_cte`.

2. The median calculation I perform here is aliased as a field named `median`.

With the CTE object wired and ready to go, I can use it in the actual jOOQ query like so:

```
context.with(medianCte) (1)
                        .select(field("brand"), field("model"),
                        field("trim"),field("price"),
concat(round((price.subtract(medianCte.field("median"))).
divide(2),2).multiply(100),field("'%'"))
                            .as("relative to median price"))
                        .from(completeCarListing, medianCte) (2)
                        .fetchMany();
```

It gives me a much less cumbersome main query:

1. I just drop in my CTE into the `with` method on `DSLContext`.

    - Take note of my usage of `medianCte.field("median")`. For all intents and purposes, the CTE will be treated almost like a table after the `WITH` clause. As a result, I can refer to (or "dereference") the columns available in my CTE just like I would any SQL table or subquery.

2. Then I can use the CTE like any old table.

Pop Quiz: What is the jOOQ parent class of all the clauses and smaller bits that can be composed into a full SQL statement ?

a) {`Query`}

b) {`QueryPart`}

c) {`Field`}

Answer: {`QueryPart`} is the parent class that can be used to represent every part of your SQL statement. You can compose any kind of SQL statement with all the children of {`Query Part`}

With the GROUP BY DSL

Grouping query results in jOOQ is as straightforward as anything else you've seen so far. I've also been quietly using the `groupBy` clause without introducing you two. No more! I'd now like to retrieve

- All Sedans
    - Grouped by brand
- The grand total of all the Sedans regardless of brand

In MySQL, the query for this would look something like the following:

```
select brand, count(*) "# of units"
from complete_car_listing
where brand = 'Sedan'
group by brand with rollup
```

Oracle uses the `GROUP BY ROLLUP (...)` syntax. Translating that to jOOQ is the same syntax:

```
Result<Record2<Object, Integer>> results = context.
select(field("brand"),count().as("units"))
                    .from(table("complete_car_listing"))
                    .groupBy(rollup(field("brand")))
        .fetch()
```

The trusty old `org.jooq.DSL` supplies all the grouping functions I'll need to pull this query off – the `count` and `rollup` functions come from there. The other grouping functions (`cube`, `groupingSets`) are also on the menu. Bon appétit!

With the HAVING DSL

With or without the `GROUP BY` clause, you can *have* the `HAVING` clause to restrict groups – in SQL and in jOOQ. To restrict my list of vehicle counts to brands with an average price higher than $20,000, I would have a jOOQ query that looks like this:

```
Result<Record2<Object, Integer>> results = context.
select(field("brand"),count().as("units"))
                    .from(table("complete_car_listing"))
                    .groupBy(rollup(field("brand")))
                    .having(avg(field("price",BigDecimal.
                    class)).gt(BigDecimal.valueOf(20000L)))
        .fetch()
```

That's it. Nothing fancy to it.

With the ORDER BY DSL

Ordering query results with jOOQ is also as straightforward as you can imagine. I've used it a fair bit already in many queries up to this point to order my car search results; I simply add the `orderBy` clause to the fluent chain I've constructed:

```
List<CompleteVehicleRecord> allVehicles = context.select()
      .from(table("complete_car_listing"))
      .orderBy(field("year").desc() ,two())
      .fetchInto(CompleteVehicleRecord.class);
```

Like I've been doing, I convert my plain SQL year column into an instance of Field. I then call the desc method on the Field instance to convert it to an OrderField – a Field type dedicated to ordering query results. Additionally, I use the two method to pass the literal "2" to the ORDER BY clause. This will additionally order the query results by the second column in the result set. Also note that you can supply a list of sort fields to the orderBy method.

ORDER BY CASE

One underrated approach to ordering is being able to conditionally order query results. If you think of "pinned" posts on forums like Reddit or "sticky" articles on blogs, this is one way to achieve that. This is a mechanism that will ensure that specific rows of result set will be positioned within the results based on specific criteria. For Eden's car shop, I want to be able to permanently list "featured" vehicles that will be at the top of every search result page. With SQL, that could look like

```
SELECT *
FROM edens_car.complete_car_listing
ORDER BY CASE featured
            WHEN true then 0
         ELSE 1 END ASC
```

Representing this in jOOQ will look something like

```
List<CompleteVehicleRecord> allVehicles = context
                    .select(field(name("brand")),
                    field("model"), field("price"))
                    .from(table("complete_car_listing"))
```

```
                    .orderBy(
                            case_(field("featured"))
                                    .when(true,0)
                                    .otherwise(1))
                    .fetchInto(CompleteVehicleRecord.class);
```

What I've done here is to

1. Specify that I want vehicles with `weight` set to `-1` listed first.
2. Then I want any other arbitrary sort criteria passed in dynamically applied.

This will guarantee that vehicles that are flagged as "featured" always show up at the top of any search results. Another approach to exerting more control over ordering is to use a map of my sort criteria to manually determine the positioning of specific rows in the overall query result:

```
HashMap<String, Integer> sortMap = new HashMap<>();
sortMap.put("Toyota",Integer.valueOf(0));
sortMap.put("Acura",Integer.valueOf(5));
```

The preceding map indicates Toyotas should come first in the list of results; Acuras should start from position 6 in the list; rows weighted 2 should show up around the 4th row. I can then pass the map to the `orderBy` clause like so:

```
.select(field(name("brand")), field("model"), field("price"))
                    .from(table("complete_car_listing"))
                    .orderBy(field(name("brand"),String.class).
                    sort(sortMap))
                    .fetchInto(CompleteVehicleRecord.class);
```

Alternatively, I could use the...

ORDER BY NULL

The behavior of the `ORDER BY` clause varies from database to database. Oracle and PostgreSQL will treat nulls as larger than others, so that when you `ORDER BY weight ASC`, rows with nulls show up last. MySQL, SQL Server, and SQLite treat nulls the opposite way: those rows will show up first when you `ORDER BY ASC`. Being specific about the direction in which you want nulls sorted is exactly the sort of deliberate coding you'll want to do if you like consistent results across RDBMSes. So, consider this SQL query:

```
SELECT *
FROM edens_car.complete_car_listing
ORDER BY color NULLS FIRST
```

Now, MySQL doesn't support this syntax; PostgreSQL and a couple others do. For MySQL however, nulls are considered to weigh the least. So, nulls will show up first when you sort in ascending order. jOOQ will transparently emulate this function so it's available regardless of the underlying server. To represent the same results in jOOQ with the orderBy DSL, I'll have

```
List<CompleteVehicleRecord> allVehicles = context
                    .select(field(name("brand")),
                    field("model"), field("price"))
                    .from(table("complete_car_listing"))
                    .orderBy(field(name("trim"),String.class).
                    asc().nullsLast())
                    .fetchInto(CompleteVehicleRecord.class);
Easy peasy. On to the next!
```

Paginate Query Results

There's the standard SQL `OFFSET...FETCH` clause to restrict the number of items returned from a query and to paginate results. MySQL's dialect provides the `LIMIT...OFFSET` clause that performs largely the same function. MySQL, H2, PostgreSQL, SQLite, and HSQLDB all support LIMIT keyword. Oracle and a couple other databases are in the OFFSET...FETCH camp. jOOQ's going to make the difference between the two syntaxes irrelevant. A query to retrieve the top 10 most expensive vehicles in MySQL will look like the following:

```
SELECT *
FROM complete_vehicle_listing
ORDER BY price DESC
LIMIT 10
```

The same query will be written in Oracle like this:

```
SELECT *
FROM complete_vehicle_listing
ORDER BY price DESC
FETCH NEXT 10 ROWS ONLY;
```

Either database will seamlessly get the correct SQL interpretation in jOOQ with the following snippet:

```
context.select()
      .from(table("complete_car_listing"))
      .orderBy(field("price").desc())
      .limit(10)
      .fetchInto(CompleteVehicleRecord.class);
```

The limit method is all I need. I can add the WITH TIES SQL clause to ensure that within with my top-N query, rows that have the same value for the ORDER BY column (i.e., tied) will be featured in the results. What this means is that when I query for the top 10 most expensive vehicles in the database, if vehicle #11 is tied with #10 for price, it will be included in the query results, regardless of exceeding the limit clause. Here's what that looks like in jOOQ:

```
context.select(fieldList)
        .from(table("complete_car_listing"))
        .orderBy(field("price").desc())
```

```
                    .limit(10)
                    .withTies()
                    .fetchInto(CompleteVehicleRecord.
                    class);
        logger.info(allVehicles.toString());
```

To get a sorted list of vehicles in descending order of price and to support pagination, I'll introduce the `offset` clause in my jOOQ DSL:

```
<jooq offset>
```

This query will retrieve a page of `value` results, with an `offset` starting point. Note that you'll use the `limit` function regardless of the type of RDBMS you're using - jOOQ handles the translation.

Pop Quiz: Which jOOQ class lets you construct conditional `WHERE` clauses?

Answer: The `DSL` class, with the `DSL#noCondition()` function.

Insert Statements

The veritable "C" in CRUD. Inserting one new vehicle into the database with jOOQ looks like this:

```
context.insertInto(
table("vehicle"),
        field("vehicle_manufacturer"),field("vehicle_curr_
        price"),field("vehicle_model_year"),field(
        "vehicle_status"),field("vehicle_color"),
        field("vehicle_model_id"),field("vehicle_trim"),
        field("vehicle_style"),field("featured"))
```

```
.values(4,BigDecimal.valueOf(46350.00), LocalDate.
parse("2021-01-01").getYear(),"ACTIVE","BLUE",
13,2,1,1)
.execute();
```

The insertInto node provides the fluent API to deliver on the features that we've come to expect from jOOQ. For all incarnations of the insert API, the first argument is a Table; after that, you can optionally provide the individual fields to insert as is the standard SQL INSERT statement.

In addition to the standard INSERT...VALUES syntax, jOOQ offers some other flavors, such as...

With Multiple Rows

I can insert multiple vehicles into the database like so:

```
context.insertInto(table("vehicle"),field("vehicle_manufacturer"),
field("vehicle_curr_price"),field("vehicle_model_year"),
field("vehicle_status"),field("vehicle_color"),field("vehicle_
model_id"),field("vehicle_trim"),field("vehicle_style"),
field("featured"))
                        .values(4,BigDecimal.valueOf(46350.00),
                        LocalDate.parse("2021-01-01").getYear(),
                        "ACTIVE","BLUE",13,2,1,1)
                        .values(9,BigDecimal.valueOf(83000.00),
                        LocalDate.parse("2021-01-01").getYear(),
                        "ACTIVE","GREY",20,9,1,1)
                        .values(9,BigDecimal.valueOf(77000.00),
                        LocalDate.parse("2016-01-01").getYear(),
                        "ACTIVE","WHITE",20,9,1,1)
                    .execute();
```

Caution While the multi-values insert is part of the ANSI standard SQL specification, it's not uniformly supported by all databases. jOOQ will emulate this for noncompliant databases. Even then, you might still hit a wall as a result of the maximum packet size that's allowed by the database server. For MySQL, this is the `max_allowed_packet` server parameter.

With Sequences

While MySQL provides the `AUTO_INCREMENT` function to autogenerated indices, you can still get a hold of custom sequences and trigger a generation this way:

```
BigInteger nextVehicleManufacturerId = context.nextval
("vehicle_manuf_seq");
```

Sequences are a lot more fun to use when generated for you by jOOQ though, so stay tuned!

With Select

I can use the `INSERT...SELECT` standard SQL syntax to copy rows from one table into another. For my use case, I'll use this syntax to archive vehicles that have been sitting in the inventory for a long time. Using the age of the row as the filter condition, I've determined I want to copy vehicles from `vehicles` to `vehicle_archive`. The way this will look in jOOQ:

```
context.insertInto(table("vehicle_archive"))
                    .select(DSL.selectFrom("vehicle").where
                    ("datediff(date(now()),created) < 365"))
                   .execute();
```

...and that's it. Moving on!

Update Statements

Yes, I too have accidentally done an UPDATE...SET without the WHERE clause. I'd rather not talk about how much destruction I wrought as a result.

And lady, if you tell anyone who set the status column to the same for all 500k rows...

Here's what it'll look like when I want to update the price of an existing vehicle in the inventory:

```
context.update(table("vehicle"))
                    .set(field("featured"),false)
                    .where(field("vehicle_id").eq(7))
                    .execute();
```

Also very straightforward. Now, jOOQ can protect you (i.e., me) from accidental UPDATE... without a where clause with the setExecuteUpdateWithoutWhere method on the Settings class.

```
//Non! Disallow updates without a where clause by throwing an
exception
new Settings().setExecuteUpdateWithoutWhere(ExecuteWithout
Where.THROW);
```

Together with the `ExecuteUpdateWithoutWhere` enum, you can configure whether to

- Ignore the condition with `IGNORE`
- Log a warning with `WARN`
- Log at debug with `DEBUG`
- Log at info level with `INFO`
- Fail all attempts to do this with `THROW`

No more messes to clean up!

Delete Statements

It really is time to move on. I've archived the vehicles I couldn't sell. Now's the time to get rid of them from the inventory altogether.

One of you will cease to exist shortly. Say your goodbyes

Pretty straightforward, the delete DSL (because, of course it is):

```
context.deleteFrom(table("vehicle_archive"))
                    .where(field("vehicle_id").eq(7))
                    .execute();
```

Simple, yes? Great. Now let's try...

Tuple Syntax

I can get a little bit fancier with my delete statement. As you may have noticed already, jOOQ supports the tuple syntax (a.k.a. row value expressions) where we can do whole row comparisons:

```
UPDATE vehicle_archive
SET (vehicle_status,featured) = ("ARCHIVE",0)
WHERE (vehicle_status,featured) = ("ACTIVE",1)
```

It's basically hard coding, but for SQL. With this query, I'm asking the query to delete all vehicles, except the specific rows or tuples that match the specific combination of columns that I've specified. This way, I update everything except these specific rows or tuples.

MySQL doesn't support this. Yes, I could have also written this as separate clauses in the WHERE condition, but where's the fun in that?

To replicate this in jOOQ terms, I'll have

```
context.update(table("vehicle_archive"))
                    .set(row(field("featured"), field("vehicle_
                    status")),
                            row(1,"UNARCHIVED"))
                    .where(row(field("vehicle_status"),
                    field("featured")).eq(row("ACTIVE",0)))
                    .execute();
```

Goodbye to those vehicles (except the ones I've chosen to save for some reason)!

Alternative Data Access Modes

I've shown only synchronous data access operations so far:

- A user request initiates the CRUD.
- The calling thread waits for the data to return from the database.
- There's some transformation work done in the same calling thread.
- Then the data is returned to the caller.

All of this happens in one thread. The `org.jooq.Result` class, the fundamental unit of handling database query results, contains *all* the results returned from the query. This has the benefit of not needing an open database connection to access all your results. The downside is that for large query results, you'll be using up a lot of memory. There's also the added disadvantage of single threading the processing of large results.[12]

<glutton image>

But there are other ways – the ways of the lazy, the streamer, and the reactive. Let's first talk about streaming.

[12] Careful here: there's a point of diminishing returns when multithreading over data. I talk about this in my concurrent data access course on LinkedIn: `www.linkedin.com/learning/java-concurrency-troubleshooting-data-access-and-consistency/java-streams-and-lambda-concurrency-issues`

Streaming Access

jOOQ offers a couple of conveniences for streaming data from the database, in every sense of the word "stream":

- The `org.jooq.Result` class extends `java.util.List`. Therefore, you have access to all the features that `java.util.Stream` affords you simply by opening a stream on your instance of `Result` after a query.
- You can call the `fetchStream` method on the `Result` class as a convenience for the same purpose.

Before we go any further, I should be clear: using the `stream`-centric jOOQ functionality changes the operating model somewhat. Where `org.jooq.Result` will load all your results into memory and disconnect from the open database connection, `fetchStream` will sustain the open database connection. What this means is that you will now need to remember to close the connection when you're done processing. So, what does stream code in jOOQ look like? Remember how I like deals on cars?

```
DSLContext context = DSL.using(connection,
SQLDialect.MYSQL);
try(final Stream<Record> records = context.
select().from(table("complete_car_listing"))
            .fetchSize(100) (1)
            .fetchStream()){ (2)
    records.parallel().forEach(recordList -> {
       //deal with records
    });
}
```

In a change from how I ran this same query earlier, I'm now running it inside a `try...with resources` block. This means that all associated underlying resources will be closed automatically after I exit the block. Under the hood, jOOQ uses an `org.jooq.Cursor` to efficiently process the results in chunks from the database.

1. I use `fetchSize` setting to hint Connector/J (the MySQL driver) that I want my query results streamed row by row instead of loading it all into memory at. Yes, this is in addition to jOOQ's own best effort attempts to do the same. I'm highlighting this specifically because at this point, different databases will give you different behaviors when you attempt to stream with Cursors.

    - Because I'm using MySQL, I **must** finish consuming all the results associated with this query,[13] on the instance of the JDBC connection that I'm using to serve the results. Failing to do so will render the connection useless for any other thread in the same application – which is asking for trouble in a connection pool scenario.
    - Any locks associated with the rows in the result will be held until all the rows are read.
    - As a result of all this, I want to get through the query results as quickly as reasonably possible.

2. Then I parallelize the stream so that I can use a couple of threads to run through the results faster.

[13] `https://dev.mysql.com/doc/connector-j/8.0/en/connector-j-reference-implementation-notes.html`

Note Standard Java stream rules still apply. For example, once I exit the `try` block, the stream is no longer accessible; neither can you scroll backward in a stream – once an item is consumed, it's done.

This is one way to handle data, but it's still fairly synchronous. What else is there?

Lazy Access

If it starts with "lazy," I'm already halfway there. jOOQ provides the `fetchLazy` method as the One True Way to properly fetch data in manageable chunks. It's usable for when you don't need the conversion of `org.jooq.Result` to a stream. This time, I'll need to handle the `org.jooq.Cursor` myself:

```
DSLContext context = DSL.using(connection, SQLDialect.MYSQL);
            try(final Cursor<Record> records = context.
            select().from(table("complete_car_listing")).
            fetchSize(100).fetchLazy()){ (1)
                    while(records.hasNext()){ (2)
                        CompleteVehicleRecord
                        completeVehicleRecord = records.fetch
                        NextInto(CompleteVehicleRecord.class);
                    }
            }
```

1. A cursor is a resource, so I'm still opening it in a `try-with-resources` block.
2. I work through the items as normal.

The same caveats apply as in `fetchLazy` stream: the `Cursor` maintains an open JDBC connection and `PreparedStatement`, so don't keep it around longer than it needs to be. This and `fetchStream` are the best bet for running large queries.

Transactions

They are the fundamental unit of every SQL operation carried out in the database. Yes, transactions are happening in the database whether you explicitly do anything to define them or not. This section is about being deliberate about transaction settings when operating in jOOQ. I'm going to show you how to use jOOQ to deliver on the ACID guarantee. ACID stands for

- **A**tomicity: When you designate a block of code as being transactional, any failure in execution or exception thrown within that block will cause the reversal of all the Database Manipulation Language (DML) executions within that block. This means all inserts, updates, or deletes.
- **C**onsistency: Means that any transactional code block that executes DML changes is guaranteed to adhere to any integrity rules defined in the underlying datasource. So that uniqueness, constraints, foreign-primary key relations will be respected by any attempts to modify the data.
- **I**solation: Database transactions can be configured to protect the data being operated on from multithreading-related corruption. Some of the issues to protect against include
    - Dirty Reads: Where one thread can read uncommitted data being written by a different thread

- Non-repeatable Reads: Where one thread reading from the same row within a time window will get different results each time
- Phantom Reads: Where data essentially disappears during successive reads by the same transaction
- Lost Updates: Where two transactions (threads) executing updates against the same row corrupt the data, each without knowledge of the other's actions

- **D**urability: The warranty that changes that have been written to the database and acknowledged by the RDBMS are guaranteed to be persisted and retrievable.

So, that's ACID. How does this work in jOOQ? Plain jOOQ, without Spring, JTA, or any other framework with dedicated transaction management, will defer to standard JDBC semantics. This means manually defining the transaction boundaries like so:

```
context.transaction(configuration -> { (1)
                updateVehiclePrice(configuration);
                insertPriceChange(configuration);    (2)
                configuration.dsl().transaction(innerConfig ->{
                    //more work
                });
            //profit??     (4)
            });
```

1. This line marks the start of a transaction boundary. Here:
   - jOOQ sets `AUTO_COMMIT` off for the underlying JDBC driver.
   - It supplies an instance of `org.jooq.TransactionalRunnable` (no, not that `Runnable` from `java.lang.Runnable`). This is my handle to begin executing in my transaction boundary.

2. My `insert` and `update` execute as normal, but without being committed. This way, if either of them fail for any reason, the entire code block is unwound with the exception that caused the failure.

    - Note how I'm passing the `Configuration` object into the nested methods. This is necessary to ensure the database operations in those methods participate in the same transaction boundary.

3. Additionally, I can nest transactions so that

    - This transaction block inherits from the outer transaction block.
    - A failure within this block will roll the operation back to the last save point before this block. This way, the rest of the operation can continue if that's what I choose.

4. If everything proceeds without choking on an exception up till this point, another transaction boundary is defined. This then marks the whole transaction as complete.

This is the way of the default jOOQ transaction provider. You have the option of supplying a custom transaction provider in the way of the Spring framework or others; that comes later in this book.

Caution This approach to transaction handling delivers on only the Atomicity component of the ACID guarantee. Most RDBMSes will offer Consistency and Durability out of the box, perhaps with some tuning. To protect against lost updates and other Isolation-related problems, you'll need to do a little bit more work – still within the jOOQ framework.

With Locking

Locking is how you get the Isolation part of the ACID guarantee. Specifically, you're able to

- Support concurrent reads of table rows
- Prevent Isolation-related failures by causing concurrent updates and deletes to fail

What this means in jOOQ is that you can be sure that when two transactions (or threads) are trying to modify a row concurrently, only one of them will succeed. The other will get an `org.jooq.exception.DataChangedException` when it tries to commit its changes.

There are different approaches to supporting locking:

- `SELECT FOR UPDATE` is a SQL standard query that locks rows in preparation for an update or a delete operation. jOOQ will transparently run this query before executing the actual DML. This is what's known as pessimistic locking.
- Using Multi-Version Concurrency Control (MVCC), a form of what's known as optimistic locking. This approach is supplied by jOOQ only when you use jOOQ-generated code; more on this approach later in the book.

The MVCC approach is available only with jOOQ-generated code. For plain SQL, you can have the pessimistic locking like so:

```
context.select(field("vehicle_curr_price"))
          .from(table("vehicle")
          .where(field("vehicle_id").eq(11))
          .forUpdate()
          .fetch();
```

The `forUpdate` call in the preceding snippet will obtain an exclusive lock to the affected row in the underlying table. This means that no other database transaction (or application thread) can perform DML on that row. With MySQL, you could even achieve blocking any other thread from reading the same row. The underlying mechanism that supports this varies from database to database.

Now, locking is all well and good in happy path scenarios. This kind of power could become problematic when a lock doesn't get released due to application error or bad weather on a given day. Like any locking mechanism in software engineering, you typically want some form of a failsafe. I have a couple of options:

1. The `wait` method lets me specify a timeout for either attempting to acquire a row lock or holding on to an existing lock. This way, I don't inadvertently wait forever to acquire a lock:

   ```
   context.select(field("vehicle_curr_price"))
               .from(table("vehicle")
               .where(field("vehicle_id").eq(11))
               .forUpdate().wait(3000)
               .fetch();
   ```

   It accepts the timeout in milliseconds. This syntax is supported by MySQL, Postgres, Oracle, and MSSQL. The default behavior is to not wait at all for a lock.

2. MySQL[14] and Postgres offer a `forShare` clause as an enhancement that supports shareable locks. This way, other threads can still read the same row, while the thread that owns the lock can commit changes:

[14] Only the InnoDB engine in MySQL provides locking.

```
context.select(field("vehicle_curr_price"))
          .from(table("vehicle")
          .where(field("vehicle_id").eq(11))
          .forShare().wait(3000)
          .fetch();
```

forShare also supports the wait flag.

3. jOOQ also supports the SKIP LOCKED option with... skipLocked:

```
context.select(field("vehicle_curr_price"))
          .from(table("vehicle")
          .where(field("vehicle_id").eq(11))
          .forShare().skipLocked().wait(3000)
          .fetch();
```

skipLocked will make the transaction ignore rows or tables that have already been locked by another transaction. The safety valve wait is available here as well.

Overall, you want to be cautious with pessimistic locking. If your transaction isn't committed, your calling thread hangs on to the lock and we're all going to have a bad time.

Configuration

Let's try for some more advanced control of the jOOQ runtime. There are a couple of interesting attachments you can plug into the runtime to gain more control over how jOOQ works. Check these out.

Connection Management

I've been using a solitary JDBC connection for my examples thus far:

```
try (Connection connection = DriverManager.
getConnection("jdbc:mysql://localhost/edens_car?user=test&pass
word=thisisabadpassword")) {
    //business things
}
```

In a production-grade deployment, you need something more...production-y. jOOQ provides an `org.jooq.impl.DefaultConnectionProvider` to handle the default usage of jOOQ – a single connection that you supply to the context like I show in the preceding snippet. In a production strength application, you're more than likely going to be dealing with a connection pool and an instance of `javax.sql.DataSource`. What to do?

Enter the `org.jooq.ConnectionProvider` interface. This is an extension point you can implement to take more responsibility for how connections to your database are handled. `ConnectionProvider` offers two methods:

```
public class CustomConnectionProvider implements
ConnectionProvider {
    DataSource dataSource;

    @Override
    public Connection acquire() throws DataAccessException {
        try {
            return dataSource.getConnection();
        } catch (SQLException e) {
            e.printStackTrace();
        }
        return null;
    }
```

```
    @Override
    public void release(Connection connection) throws
    DataAccessException {
            try {
                connection.close();
            } catch (SQLException e) {
                e.printStackTrace();
            }
    }
}
```

The jOOQ runtime will call `acquire` to obtain the JDBC connection for statement execution. It will then call `release` to dispose of the connection when execution is done. Here, I'm interested in tracking the interval between connection acquisition and disposal. This is a crude way to keep an eye on how long my query is executing. The disposal mechanism depends on what the underlying configuration stipulates. When you're working with a connection pool, the connection won't actually be closed; it'll be returned to the pool for subsequent reuse.

jOOQ also offers the `DataSourceConnectionProvider` as direct support for `javax.sql.DataSource`. So, in a Spring Boot application, for example, this is probably what you'll be using. To use my connection provider:

```
Configuration config = new DefaultConfiguration();
CustomConnectionProvider customConnectionProvider = new
CustomConnectionProvider();
//set it directly on the Configuration
config.set(customConnectionProvider);
//Alternative: pass it to the DSLContext
DSLContext context = DSL.using(new CustomConnectionProvider(),
SQLDialect.MYSQL,settings);
```

It's that simple: pass it to an instance of `Configuration` which in turn goes into the `DSLContext`; alternatively, pass it directly to the DSLContext. Unless you do something you're not supposed to, the `DataSourceConnectionProvider` should be thread safe. What this means is that you could design your app to reuse the same `DataSourceConnectionProvider` across the app, plugged in to the same connection pool. You can also pass the datasource directly to your configuration and skip all the other hassle.

Caution When you implement a custom `ConnectionProvider`, you'll lose access to a few convenience methods that the default jOOQ implementation provides. For example, `commit` and `rollback` are off the table. You're tacitly taking a few matters into your own hands with this feature.

Schema, Catalog, and Multi-tenant Deployment

Databases support some combinations of the following:

- Schema: A collection of related tables, views, stored procedures, and functions. It's the bag for all the "stuff" you create in the database. Not all databases see it this way – for example, MySQL considers the database and everything inside as the schema. Oracle and SQL Server consider the schema as separate from other contents of the database server.

- Catalog[15]: The catalog is a higher level of abstraction of the schema. So, a catalog can contain multiple schemas (or schemata if you're fancy). MySQL thinks exclusively in terms of catalogs so that there's no schema - the catalog is the collection of related tables, etc.

These two mechanisms can be used to support multiple separate instances of a single application on the same database server instance. This means that for different clients of your application, they can share the same database server instance(s) with their tables walled off and separate. This is what's known by the nerds as multi-tenancy.

Given that I'm using a MySQL database for my online car sales website, I have to use the catalog as the discriminator to support multi-tenancy. When you're using jOOQ-generated code, jOOQ by default will qualify all components (tables, sequences, views, etc.) with the schema/catalog name to be sure we're routing all queries to the correct schema. You can disable that with the following setting:

```
new Settings()
  .withRenderCatalog(false)
  .withRenderSchema(false);
```

This way, you can control the schema or catalog used at runtime by simply prefixing the elements of your SQL query with the correct catalog.

What if you need to do this at runtime? You have a "master" or "dev" schema that you're working on at development time. At deployment time, you would want "master" or "dev" translated to a production schema or even a dynamic schema specified at runtime. What does that look like in jOOQ terms?

---

[15] As of jOOQ 3.15, catalogs are available only for Sybase ASE and SQL Server.

```
Settings settings = new Settings()
            .withRenderMapping(new RenderMapping()
            .withDefaultSchema("default_schema")
            .withDefaultCatalog("default_catalog)
            .withSchemata(new MappedSchema().
            withInput("master").withOutput(schemaInEffect))
            .withCatalogs(new MappedCatalog().
            withInput("master").
            withOutput(schemaInEffect)));
      ...
      DefaultConfiguration configuration = new
      DefaultConfiguration();
      configuration.setSQLDialect(SQLDialect.MYSQL);
      configuration.setConnection(connection)
      configuration.setSettings(settings);
      DSLContext context = DSL.using(configuration);
```

This `Settings` snippet demonstrates the usage of the `MappedSchema` and `MappedCatalog` classes.

- With these two classes, I can instruct the jOOQ runtime to translate an input schema (`master` in this case) to a different runtime schema.
- The `withDefaultSchema` and `withDefaultCatalog` methods let me set up a default schema for all queries. These schemas **will not** be used as prefixes for any query components – it's the default, so there's no need to specify them in every query.
- I then feed the enclosing `Settings` instance into a `Configuration` instance. The configuration can then be fed to the `DSLContext`. This means that my query components (tables, sequences, stored procedures,

etc.) will be prefixed with an `org.jooq.Catalog` or `org.jooq.Schema` class that specifies which catalog or schema I want to deal with at runtime.

Another approach to dynamically configuring the schema is to set it on the `DSLContext` itself with

```
context.setSchema(selectedSchema).execute();
```

//or

```
context.setCatalog(selectedCatalog).execute();
```

where `selectedCatalog` or `selectedSchema` could be an `org.jooq.Name`, a plain string, or instances of `org.jooq.Catalog` and `org.jooq.Schema`, respectively. For objects in your query that don't already have a schema prefix, this sets the active schema or catalog at runtime.

Query parts for the win!

Query Management

There are more than a few facilities to control how SQL statements are handled by the jOOQ runtime. Feast your eyes.

- Pretty print SQL with `withRenderFormatted`:

  ```
  new Settings().setRenderFormatted(true);
  ```

- Control how identifiers are rendered. Different databases provide some perks depending on whether the identifier is quoted or not.

  ```
  .withRenderQuotedNames(RenderQuotedNames.ALWAYS);
  ```

  The `RenderQuotedNames` enum provides a few options to control this behavior:

- Optimize the performance of SQL statements that use the IN comparison parameter. Because of the way most databases cache PreparedStatements, it's important that IN comparisons use a constant-sized list of items. What this means is that this statement

  ```
  SELECT * from complete_vehicle_listing where vehicle_
  manufacturer in (?)
  ```

  and this statement

  ```
  SELECT * from complete_vehicle_listing where vehicle_
  manufacturer in (?,?,?,?)
  ```

  will be treated as different statements by the database, even though the only difference is that the number of parameters in the IN list is different. In a high traffic RDBMS, this can yield shockingly poor performance. jOOQ's solution is called "IN list padding." With this feature, jOOQ will pad the query parameter with a constant multiplier. This will help increase the rate at which the database can cache PreparedStatements, yielding better performance. It's a simple Settings operation:

  ```
  new Settings().withInListPadding(true) //defaults to
                                                    false
               .withInListPadBase(4) //starting count to
                                                  pad with
  ```

- Set JDBC parameters, for example, queryTimeout and maxRows:

  ```
  Settings settings = new Settings().withQueryTimeout(5)
  //in seconds
        .withMaxRows(1000)
  ```

Check out the Settings javadoc for more interesting parameters you can tweak.

Query Lifecycle Integration

I'll use jOOQ's org.jooq.ExecuteListener to key into the query execution by the jOOQ runtime. It provides the opportunity to intercept the process during up to 20 events. jOOQ ships with these two listeners, among others:

- org.jooq.tools.StopWatchListener is a listener that will help you track the execution times around key events in the query execution process.
- org.jooq.tools.LoggerListener prints log statements during key events of the statement execution lifecycle.

ExecuteListeners are ideal for introducing cross-cutting functions like logging (like you've already seen), exception handling, or even security features.

```
public class QueryIntrospectionListener extends
DefaultExecuteListener { (1)

    final Logger logger = LoggerFactory.getLogger(Query
    IntrospectionListener.class);

    @Override
    public void fetchStart(ExecuteContext ctx) { (2)
        logger.info("Executing: {}",ctx.sql()); (a)
        logger.info("Query type: {}",ctx.type()); (b)
        //ctx.query().getBindValues() to retrieve bind
        parameters for the query
    }
```

```
    @Override
    public void exception(ExecuteContext ctx) { (3)
        if(Objects.nonNull(ctx.sqlException())){
            //handle exception
        }
    }

}
```

First off, I should mention that I'm a big fan of the context object[16] pattern, and all my favorite frameworks lean into it, hard. Here's what's going on with the preceding snippet:

1. Extending `org.jooq.impl.DefaultExecuteListener` is the recommended approach to getting your own listener going. That class contains many lifecycle methods that you can override. Overriding these methods provides privileged access to the jOOQ runtime, so you can inject your own logic and designs into the overall query execution process.
2. I've chosen to override the `fetchStart` method. This means that I can step in before the execution of the `fetch` operation of the jOOQ API. This method (and all the others in `DefaultExecuteListener`) is supplied with an instance of `org.jooq.ExecuteContext`. This beautiful context object contains all the **contextual** information you'll need about the currently executing query. I have access to

[16] https://stackoverflow.com/questions/986865/can-you-explain-the-context-design-pattern/

a. The raw SQL being executed with

b. The type of query being executed:

c. The actual Query object and inspect the parameters.

3. In my overridden exception method, I can get a hold of any exceptions that occur during query processing and do...whatever I want with it.

With my custom listener defined, here's how I plug it into the DSLContext:

```
Configuration configuration = new DefaultConfiguration();
            configuration.set(connection)
                    .set(SQLDialect.MYSQL)
                    .set(new QueryIntrospectionListener());
            DSLContext context = DSL.using(configuration);
```

To be able to use my custom ExecutionListener

1. I'll use the set method to supply an instance of that listener to my instance of org.jooq.Configuration.

2. I then use that configuration to obtain a DSLContext and voila!

Being able to interweave custom logic into the execution of queries and the richness of context presents so many opportunities for customization.

Man, I love context objects.

CHAPTER 3

Working with jOOQ

And now for my next demonstration, we're going to take a much deeper dive into the jOOQ toolbox. There's *a lot* of convenience in the jOOQ toolbox – heck, jOOQ is all about convenience when you think about it. All of the hand-rolled SQL I was doing in the last chapter, the hand-made entity classes and parsing of query results; it can get significantly easier and more importantly typesafe. And that's the second key to enjoying jOOQ: type safety. Put together, this chapter is a saunter through jOOQ's features that offer convenience and type safety. First, let's look at jOOQ's code generation capabilities.

Generating Code

This is approximately half the point of jOOQ as a concept: being able to typesafely refer to columns, tables, sequences, really, any part of a SQL query. Among other advantages, you'll find that

- Incorrect SQL will no longer be a thing to worry about when the SQL is derived directly from what is in your database; zero guesswork required.
- A lot of boilerplate code will be automatically taken care of. I can tell you from experience that it's nice to not have to type out yet another entity class or Data Access Object (DAO) by hand.

T. Koleoso, *Beginning jOOQ*, https://doi.org/10.1007/978-1-4842-7431-6_3

- Your IDE experience is markedly improved by the fact that you can take advantage of things like code completion, hints, and "find usages" while working with database components.
- Your data-driven unit and integration tests become much more of a guarantee of the behavior and quality of your code. In a packaging model where your data access components (schemas, entities, etc.) are managed separately from the business logic, you can structure your application to independently validate any updates that have been made to the schema without going through a full deployment cycle. More on this technique later.

All told, typesafe generated database artifacts is where it's at, so let's go there!

Tools of jOOQ Code Generation

What we're interested in at this point is a way to use jOOQ to generate Java classes to represent the content of Eden Auto Mart database. There are three ways we can go about this:

1. Programmatically: jOOQ offers a simple API that you can use to generate classes from your database schema.
2. Command Line: You can also generate artifacts using a command-line interface.

3. Build Tools: jOOQ ships with Maven and Gradle tooling[1] that you can use to run the code generation operation.

The results from the preceding three methods are identical: java classes in packages, representing your database catalog or schema. They also share common configuration elements. Foundationally, the configuration for the code generator is represented as an XML document. The three modes of generating jOOQ code invariably wind up as this XML entity at some point in the lifecycle, mostly as a JAXB-annotated class. Therefore, it makes sense to start by looking at what that XML configuration document might look like.

Configure jOOQ for Code Generation

So, what *does* configuration look like? Hang tight: it's a fair bit of XML:

```
<?xml version="1.0" encoding="UTF-8" standalone="yes"?>
<configuration>
    <!-- Configure the database connection here -->
    <jdbc>
        <driver>com.mysql.cj.jdbc.Driver</driver>
        <url>jdbc:mysql://localhost/edens_car</url>
        <user>eden_admin</user>
        <password>_*thisisabadpassword*_</password>
    </jdbc>
    <generator>
        <name>org.jooq.codegen.JavaGenerator</name>
        <database>
            <name>org.jooq.meta.mysql.MySQLDatabase</name>
```

[1] https://github.com/etiennestuder/gradle-jooq-plugin

```
            <inputSchema>edens_car</inputSchema>
            <includeTables>true</includeTables>
            <includes>.*</includes>
        </database>
        <generate>
            <javaTimeTypes>true</javaTimeTypes>
             <daos>true</daos>
            <pojos>true</pojos>
            <pojosAsJavaRecordClasses>true</pojosAsJavaRecord
            Classes>
           <pojosEqualsAndHashCode>true</pojosEqualsAnd
           HashCode>
            <pojosToString>true</pojosToString>
        </generate>
        <target>
            <packageName>com.apress.jooq.generated
            </packageName>
            <directory>C:\Users\SIGINT-X\eclipse-workspace\
            jooq-demo\src\main\java</directory>
        </target>
    </generator>
</configuration>
```

Okay, you don't need to take it all in right now. Feel free to copy paste what's in here as is. Let me talk you through the most salient bits. After the top-level `<configuration>` tag (and its associated schema document link), there's a mix of optional and mandatory elements that control the behavior of jOOQ's code generation:

1. Database Connection: The `<jdbc/>` element defines how the code generator will have access to the database to begin with. Can't generate classes from a database you can't get access to.

2. Generator Semantics: The `<generator/>` element is where I provide additional context to the code generator. This node is a container for some other higher-level concerns. Here, I've defined

    a) The style of code I'm interested in generating – Java. Other options include Scala and Kotlin. The `<name/>` element controls the type of code generation that will occur.

    b) The database dialect I'm interested in with `<database/>`.

    c) The schema in the database that contains the artifacts from which I want to generate my classes. I also have the option to supply multiple schemas by nesting a `<schemata>` element here containing each `<schema/>` I want to generate from.

    d) The `<includeXXX/>` tags allow me to specify what **types** of components I want to generate from the database. Without this, you're more than likely going to wind up with a bunch of system components and other cruft you don't really need.

        i. `<includes/>` lets me specify **by name** what artifacts I want to include in the generation. This is the difference between saying "I want to include all tables in the generated code" with `<includeTables/>` vs. saying "I want to include these specific tables for code generation" with `<includes>`.

e) I prefer the use of the newer time classes in the `java.time` package. This way, I can use `LocalDateTime` and others instead of `java.sql.Time`, etc.

f) With `<dao>true</dao>`, jOOQ will generate Data Access Objects[2] (DAO) for each table. This means I have yet another convenient component to run typesafe queries for whole objects.

g) For actual transmission of entities in my web application, I don't want to send actual jOOQ records or `Tables` – that's not neat code. No, what I want is a POJO – a Plain Old Java Object devoid of any framework code – jOOQ's or otherwise. Enter the `<pojo/>` configuration to generate POJOs.

h) `target` helps me set the location and package structure I want my generated classes to be stored.

There's a lot more power and flexibility to configuring the jOOQ code generator; I highly recommend you check out the official manual for advanced options. For now, let's press on!

Given my preceding configuration, I should be able to use one of the available generator options to generate code from my existing database schema. What are my options?

Generate Code with Maven

jOOQ provides the `jooq-codegen-maven` Maven plugin to carry out your code generation business. The `groupId` of the plugin you use will depend on the distribution of jOOQ you're working with (commercial or open source), as well as the version of Java you're running:

[2] `www.oracle.com/java/technologies/data-access-object.html`

- Open Source Edition - `org.jooq`
- Commercial Edition - `org.jooq.pro`; `org.jooq.pro-java-8` for Java 8 support
- Free trial of the commercial version - `org.jooq.trial`

Because I'm just too cool for school and doing well for myself,[3] I'm going to pop in my commercial Maven plugin config like so:

```
<maven demo>
```

Here's what I've wrought in the preceding snippet:

1. I'm defining the use of the jOOQ plugin per standard Maven plugin usage.
2. I specify that I want the plugin to kick in during the `generate-sources` phase of the Maven build lifecycle.
3. I then declare a `generate` goal. This is the Maven goal I'll use to trigger the code generation process for the jOOQ plugin.
4. Providing the `skip.jooq.generation` property allows me to dynamically enable or disable the code generation at build time.
5. `configurationFile` points to the location of my XML config file for code generation as seen earlier. I also have the option to wholesale include the entire content of that config file in my Maven POM.xml (Maven's Project Object Model) file. It's possible, but you probably shouldn't do it because

[3] Also, Lukas hooked me up with a commercial license, gratis.

a. The code generation config file will likely see more change than the POM. Therefore, it's probably best that you cleanly separate the two to minimize the churn in the POM as a whole.

b. Separating the code generation config from your POM opens up the opportunity to version the config file.

c. It keeps your POM file shorter and therefore more readable.

d. It supports reusability. Think about it: when you have a standalone XML config file for your code generation, you can refer to that file from Maven, the command line, or even programmatically. And if you change your build system to Gradle, you don't need to rewrite a bunch of config!

With all of this in place, I can then run the following Maven command to generate the source files that I'm interested in:

```
mvn package
```

I can see all the new classes and packages in my IDE.

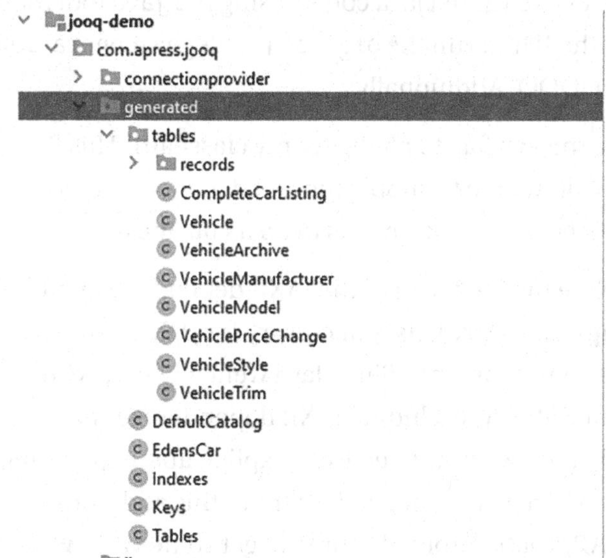

Pictured: Success

Generate Code from the Command Line

If Maven or another build tool isn't your thing, you could straight up run the generator from a terminal or command-line interface. Given the same XML configuration file, I can execute the following command from a terminal window:

```
java -cp  jooq-3.15.1.jar;jooq-meta-3.15.1.jar;jooq-codegen--
3.15.1.jar;reactive-streams-1.0.2.jar;mysql-connector-
java-8.0.24.jar;jaxb-api-2.3.1.jar ;r2dbc-spi-0.9.0.M1.jar;
org.jooq.codegen.GenerationTool jooq-configuration.xml
```

Yes, it's exactly what you're probably thinking. I'm running the code generator like it were vanilla java code.[4] Using the `java` tool that's bundled standard with the JDK, I run the `org.jooq.codegen.GenerationTool` class that ships with jOOQ. Additionally:

1. I use the `-cp` flag to configure my classpath. This flag then allows me to supply the JAR dependencies for the `GenerationTool` needs to do its business.

    a. Note the inclusion of the JAXB dependency **jaxb-api-2.3.1.jar**. This is necessary for JDK 11 and up environments. Since Java went all fancy with modularity, excluding JAXB dependencies by default, we now need to be explicit about including it whenever it's needed. Without this additional JAR, you're probably going to get some variety of `ClassNotFoundException` while running code generation from the command line. Fun.

2. I also supply the location of the XML configuration file.

I also have the option to configure the core requirements of the code generator with these fine environment variables; enjoy:

- `-Djooq.codegen.configurationFile` to define the location of the XML configuration file
- `-Djooq.codegen.jdbc.driver` to configure the driver class that will be used to connect to the database
- `-Djooq.codegen.jdbc.url` to configure the URL for connecting to the database

[4] I'm using the semicolon (";") separator here from a Windows environment; for Unix environments, go full colon (":").

- `-Djooq.codegen.jdbc.username` and `-Djooq.codegen.jdbc.password` to define the username and password, respectively, for the database connection
- `-Djooq.codegen.logging` to set the logging level for the code generation process; standard logging levels like `DEBUG`, `WARN`, `INFO`, etc. apply
- `-Djooq.codegen.skip` to disable code generation altogether

The environment variable options are a great way to set defaults for your jOOQ project. They can be overridden by what's defined in the XML file, Maven, or even in the programmatic code generation option.

Generate Code Programmatically

The most powerful option of them all [thunderclap/lightning strike]! You can programmatically generate code with the API provided by jOOQ. It goes a little something like this:

```
org.jooq.meta.jaxb.Configuration generatorConfiguration =
JAXB.unmarshal(new File("src/main/resources/jooq-
configuration.xml"), org.jooq.meta.jaxb.Configuration.class);
      GenerationTool.generate(generatorConfiguration);
```

Here, I'm simply reusing the XML configuration I previously defined, this time wrapped as an instance of `org.jooq.meta.jaxb.Configuration`. Vanilla JDK's JAXB API loads the XML file and unmarshals it to a `Configuration` object. I then use the `GenerationTool#generate` static method to run the generator.

Note The JAXB module has been separated from the JDK core; you'll need to manually include it as a dependency in your POM to be able to run the preceding sample.

For more fine-grained control, I could just do away with the XML file altogether like so:

```
org.jooq.meta.jaxb.Configuration generatorConfiguration = new
org.jooq.meta.jaxb.Configuration()
                .withJdbc(new Jdbc()
                        .withDriver("com.mysql.cj.jdbc.Driver")
                        .withUrl("jdbc:mysql://localhost/
                        edens_car")
                        .withUsername("root").withPassword
                        ("admin"))
                .withGenerator(new Generator()
                        .withName("org.jooq.codegen.
                        JavaGenerator")
                        .withDatabase(new Database()
                                .withName("org.jooq.meta.mysql.
                                MySQLDatabase")
                                .withInputSchema("edens_car")
                                .withIncludeTables(true)
                                .withIncludes(".*"))
                        .withGenerate(new Generate()
                                .withJavaTimeTypes(true)
                                .withJavaBeansGettersAnd
                                Setters(true)
                                .withDaos(true)
                                .withPojos(true)
                                           .withPojosEqualsAndH
                                           ashCode(true)
```

```
                        .withPojosToString(true))
                .withTarget(new Target()
                        .withClean(true)
                        .withDirectory("src/main/java")
                        .withEncoding(StandardCharsets.
                        UTF_8.toString())
                        .withPackageName("com.apress.
                        jooq.generated")
                ))
        .withLogging(Logging.DEBUG)
        .withOnError(OnError.LOG);
    GenerationTool.generate(generatorConfiguration);
```

This is simply a faithful duplication of the contents of the XML file for greater flexibility and more horsepower – the result is the same. You could combine the two approaches where some values could be preset in the XML file; then some values can be supplied dynamically programmatically.

Programmatic code generation is a wonderful opportunity to leverage jOOQ in scenarios where an XML file or command-line parameters won't cut it, like integration tests (more on those later). Another opportunity is to use jOOQ in slightly unconventional scenarios like when the database connection is dynamically generated.

Pop Quiz: Which element of the jOOQ configuration schema lets you configure which type of code (Java, Scala, etc.) should be generated?

Answer: The `name` element defines the output type of jOOQ code generation. Use `org.jooq.codegen.JavaGenerator` to generate Java code.

Results of Code Generation

Whichever method of code generation you choose, the results would largely be identical. Check 'em out:

Tables

Each table in your schema or catalog will largely result in the following:

1. Classes that extend `org.jooq.impl.TableImpl`, itself an implementation of `org.jooq.Table`. It'll look a little something like this for my Eden Auto database:

```
public class Vehicle extends TableImpl<VehicleRecord> {
    private static final long serialVersionUID = 1L;

    /**
     * The reference instance of <code>edens_car.
     vehicle</code>
     */
    public static final Vehicle VEHICLE = new Vehicle();

    /**
     * The class holding records for this type
     */
    @Override
    public Class<VehicleRecord> getRecordType() {
        return VehicleRecord.class;
    }
    /**
     * The column <code>edens_car.vehicle.vehicle_id
     </code>.
     */
```

```
        public final TableField<VehicleRecord, Long>
        VEHICLE_ID = createField(DSL.name("vehicle_id"),
        SQLDataType.BIGINT.nullable(false).identity(true),
        this, "");
    ...
    }
```

It's these classes that you can use to construct typesafe SQL queries; this is instead of using the `DSL#table` function that I was previously using to convert raw SQL to jOOQ types. There's no possibility of error with these classes.

2. Classes that extend `org.jooq.impl.*RecordImpl`, itself an implementation of `org.jooq.Record`. Sound familiar? `Record` is what you get back from your database queries. There are two general flavors of these that could be generated:

    a. `UpdatableRecordImpl` is what you get when the underlying table has a primary key and jOOQ has access to the primary key during code generation.

    b. `TableRecordImpl` is what you'll get when the underlying table doesn't have a primary key, jOOQ doesn't have access to the primary key data, or it isn't even a real table - a database view, for example.

    So, use your `*RecordImpl` to iterate through results of your queries; `UpdatableRecordImpl` specifically to perform DML operations against a table.

3. Plain Old Java Objects (POJOs) also come out of the box automatically, also representing rows in your database tables. Typically, a generated POJO class will extend `Serializable`. Here's what one looks like for Eden Auto:

```
public class Vehicle implements Serializable {

    private static final long serialVersionUID = 1L;

    private Long          vehicleId;
    private Long          vehicleManufacturer;
    private BigDecimal    vehicleCurrPrice;
    private LocalDate     vehicleModelYear;
    private String        vehicleStatus;
    private String        vehicleColor;
    private Long          vehicleModelId;
    private Long          vehicleTrim;
    private Long          vehicleStyle;
    private Byte          featured;
    private LocalDateTime created;

    public Vehicle() {}
  ...
}
```

Like I mentioned earlier, these come in handy as Data Transfer Objects (DTO) or value objects that you can use for shifting data around and into your application. It gets even better: you can have Bean Validation[5] specifications like `@NotNull` and `@Size` generated from information from the columns of the table. Pretty neat!

[5] https://en.wikipedia.org/wiki/Bean_Validation

4. Data Access Objects (DAOs) are analogous to Spring's various `Repository` methods. They do what they sound like: help you access strongly typed data from your tables. DAOs will help you query their respective tables to retrieve the desired records. DAOs are generated only for tables with primary keys by default. This means that a view won't automatically generate DAOs.

jOOQ allows you a great deal of control over table primary keys as part of the code generation process. If you want to add primary key information along with something like a database view, use this feature to manually inform jOOQ:

```
<database>
        <name>org.jooq.meta.mysql.MySQLDatabase</name>
        ...
        <syntheticPrimaryKeys>edens_car.complete_car_
        listing.vehicle_id</syntheticPrimaryKeys>
        ...
</database>
```

Because a view isn't really a table, most databases won't provide the same primary key information they offer for actual tables. As a result, I've had to configure the path to the key column for the database view that I'm interested. This feature is known as a synthetic primary key. With this configuration, `complete_car_listing` inside the database will produce a DAO. The downside here is that it requires some combination of

- Hard coding the name of a column
- Consistent naming conventions of primary key columns
- A regular expression that you will need to validate against the names of your primary key columns

But wait; there's more:

Global Artifacts

More convenience incoming: jOOQ can also generate most cross-cutting components as high-level "global" classes. These will be generated as static members of the following class definitions:

1. `Keys.java` will contain static fields referencing all primary, foreign, and unique keys defined on a per table basis. These will come in handy when you need to build typesafe queries with SQL joins.

2. `Sequences.java` will give you all the sequences defined in your database schema. You can reach for these when you want to manually generate a key value for some reason.

3. `Tables`.java will contain all the tables defined in your schema, useful for when you're constructing queries and such.

There are other statically generated components coming out of the jOOQ code generation process, some of which are out of the scope of this book. I highly recommend checking out the official documentation for more of the good stuff.

Pop Quiz: What jOOQ methods will produce a SQL {CASE} clause?

a) {choose}

b) {when}

c) {case}

Answer: {choose} and {when} are the valid ways to start a {CASE} statement with jOOQ

Add Custom Code to Generated Code

If you're fancy (like me), you may be interested in adding hand-woven code blocks to all (or some) of the generated code. For example, some corporate environments could be interested in adding trademark and copyright information to all their code. To pull this off, you'll need an implementation of JavaGenerator. Here's what it would look like to add a header comment to all class files:

```
import org.jooq.codegen.JavaGenerator;
import org.jooq.codegen.JavaWriter;
import org.jooq.meta.TableDefinition;

public class CopyrightGenerator extends JavaGenerator {

    protected void printClassJavadoc(JavaWriter out, String
    comment) { (1)
        out.println("/** This is proprietary code of Initech co
        */");
    }
    protected void generateRecordClassFooter(TableDefinition
    table, JavaWriter out){ (2)
        out.println();
        out.tab(1).println("public static String
        getInitTechWarning(){");
        out.tab(2).println("return \"This is proprietary code
        of Initech co\";");
        out.tab(1).println("}");
        out.println();
    }
}
```

I told you I'm fancy

This is an uber-trivial use of this class – there's almost nothing you can't rewrite or add to generated code using the `JavaGenerator` facility.

1. `printClassJavadoc` allows me to prepend any arbitrary Javadoc content to the top of a class declaration (after imports).

2. `generateRecordClassFooter` lets me append arbitrary code to the end of a `Record` class – basically any table or view. You may recognize the risk here: this facility deals with plain type-unsafe strings (ironic, I know), but there are trivial ways to make sure you never accidentally include snippets that won't compile.

The default `JavaGenerator` provides an impressive array of methods available to override and change any part of the generated code. Go ahead and have fun with it!

Working with Generated Code

When you have code that's directly woven from your database schema, type safety is a benefit in and of itself. But it doesn't stop there. See, when jOOQ is the source of your data access code, there are unique benefits that accrue.

CRUD with Generated Code

Things work better with generated code for vanilla CRUD operations. I can retrieve from the vehicle table with strong typing guarantees like so:

```
import static com.apress.jooq.generated.EdensCar.EDENS_CAR;
import static com.apress.jooq.generated.Tables.VEHICLE; (1)
...
public static void selectWithGeneratedCode() throws
SQLException {
       ...
       DSLContext context = DSL.using(connection, SQLDialect.MYSQL);
       List<Vehicle> vehicles = context.select(EDENS_CAR.
       VEHICLE.VEHICLE_ID, EDENS_CAR.VEHICLE.VEHICLE_COLOR,
       EDENS_CAR.VEHICLE.VEHICLE_CURR_PRICE) (2)
                    .from(VEHICLE)     (3)
                    .where(VEHICLE.VEHICLE_MANUFACTURER.
                    eq(val(2L))) (4)
                    .orderBy(VEHICLE.VEHICLE_MODEL_YEAR)
                    .fetchInto(Vehicle.class); (5)
            logger.info(vehicles.toString());
        }
    }
```

I've purposefully blended a couple of styles and concepts into the preceding snippet to illustrate the flexibility you get with generated artifacts.

Using the generated artifacts:

1. Static imports of the generated schema (`EDENS_CAR`) and a generated table (`VEHICLE`) classes mean that I can use their respective contents wherever I need them later.

2. I can use the fully qualified path to columns in the `select` statement - `Schema.Table.Column`.

3. I can drop the schema altogether and use just the plain generated table reference provided by the generated `com.apress.jooq.generated.Tables` class.

4. Supplying parameters for filter expressions is a breeze because

    a. I don't have to deal with placeholders like "?" or remember the order of the parameters. I can directly specify parameters on the fields that need them.

    b. Using the `val` function binds the parameter value to the correct column during the execution of the jOOQ statement; the SQL injection protection I get is a nice bonus. This isn't always necessary however - jOOQ does this under the hood in most cases.

5. Finally, I can fetch the query results into the POJO generated from the `Vehicle` table.

So far so good. I could also just do the `fetch` directly into the `UpdatableRecord` implementation for the `vehicle` table:

```
Result<VehicleRecord> vehicle = context.fetch(VEHICLE, VEHICLE.
VEHICLE_MANUFACTURER.eq(val(2, Long.TYPE)));
```

This is the concise kind of code I live for.[6] I've done away with a lot of the boilerplate; I simply provide the table class and my filter criteria. In addition to the conciseness, there's a navigation benefit to this type of fetching, and I'll be covering it shortly.

For an insert into the `vehicle` table, I can immediately retrieve the primary key associated with the insert with the `returningResult` method:

```
long execute = context.insertInto(VEHICLE, VEHICLE.VEHICLE_
MANUFACTURER, VEHICLE.VEHICLE_CURR_PRICE, VEHICLE.VEHICLE_
MODEL_YEAR, VEHICLE.VEHICLE_STATUS, VEHICLE.VEHICLE_COLOR,
VEHICLE.VEHICLE_MODEL_ID, VEHICLE.VEHICLE_TRIM, VEHICLE.
VEHICLE_STYLE)
                    .values(4L, BigDecimal.valueOf(46350.00),
                    LocalDate.parse("2021-01-01"), "ACTIVE",
                    "BLUE", 13L, 2L, 1L)
                    .returningResult(VEHICLE.VEHICLE_ID)
                    .execute();
```

The `returningResult` method allows you to return fields from the newly inserted row as part of the response from an insert. The ideal usage is to return the primary key generated for the insert as supported by the underlying database. For other non-key fields, jOOQ might perform a second SQL execution to retrieve the requested data if the underlying database doesn't natively support `INSERT...RETURNING`.

[6] This isn't a feature for just generated code; you can do this with plain SQL. It just looks cooler here. ☺

All told, the risk of getting the names of tables or columns wrong is eliminated. Combined with the fluent jOOQ DSL that makes sure that your SQL is always going to be syntactically correct: you have bulletproof database queries.

Enhancements from UpdatableRecord

With generated UpdatableRecords, you gain some powerful advantages with CRUD operations. One of my favorite bits of this component is how I can navigate the relationships of a record:[7]

```
DSLContext context = DSL.using(connection, SQLDialect.MYSQL);
        VehicleRecord vehicleRecord = context.
        fetchOne(VEHICLE, VEHICLE.VEHICLE_ID.eq(7L));
        VehicleManufacturerRecord vehicleManufacturerRecord =
        vehicleRecord.fetchParent(Keys.VEH_MANUFACTURER_
        ID); (1)
        VehicleModelRecord vehicleModelRecord =
        vehicleRecord.fetchParent(Keys.VEH_MODEL_ID); (1a)
        logger.info("Vehicle Manufacturer:
        {}",vehicleManufacturerRecord.
        getManufacturerName());
        logger.info("Vehicle Model Record:
        {}",vehicleModelRecord.getVehicleModelName());
        Result<VehicleModelRecord> vehicleModelRecords =
        vehicleManufacturerRecord.fetchChildren(Keys.
        MANUFACTURER_ID); (2)
```

[7] You do run the risk of the N+1 query issue: https://stackoverflow.com/questions/97197/what-is-the-n1-selects-problem-in-orm-object-relational-mapping

```
vehicleModelRecords.forEach(vmr ->{
    logger.info("{} {}",vehicleManufacturerRecord.
    getManufacturerName(),vmr.
    getVehicleModelName());
} );
```

After retrieving a specific vehicle, I can

1. Retrieve the manufacturer record by using the foreign key relationship between `vehicle` and `vehicle_manufacturer`. This works by using the appropriate generated foreign key from the (also generated) `Keys` class. Pass that into the `fetchParent` method and I'm sorted.

   a. I use the same mechanism to retrieve the vehicle model record associated by foreign key to the `vehicle` table.

2. I can walk down the family tree instead with `fetchChildren`, passing in the foreign key that ties `vehicle_manufacturer` and `vehicle_model_id`.

What about DML? Each `UpdatableRecordImpl` is an Active Record – this means that you don't need any additional components to perform data manipulation on retrieved records:

```
VehicleModelRecord vehicleModelRecord = context.
newRecord(VEHICLE_MODEL); (1)
vehicleModelRecord.setVehicleManId(2L);
vehicleModelRecord.setVehicleModelName("Tacoma");
vehicleModelRecord.setVehicleStyleId(3L);
vehicleModelRecord.store(); (2)    //record saved
Long vehicleModelId = vehicleModelRecord.getVehicleManId(); (3)
```

```
vehicleModelRecord.setVehicleModelName("Tacoma XD");
vehicleModelRecord.store();    (4)
vehicleModelRecord.refresh(); (5)
vehicleModelRecord.delete();    (6)
```

1. I can create an empty `VehicleModelRecord` object from the `DSLContext`. No record has been created in the database yet.

2. After setting necessary, I can use the `store` method on the object to then persist the record.

3. I can retrieve the `id` of the newly inserted record immediately.

4. I can continue calling `store` at my convenience on the same record.

5. I'll use the `refresh` method to get the latest copy of the record I'm working on.

6. And when I'm done with, I can just call the `delete` method to get remove the record.

Tip My `UpdatableRecordImpl` is *attached* to a database connection, but lazily. What this means is that my instance of `VehicleRecord` contains a reference to the `Configuration` object, which has a reference to the underlying JDBC connection pool. Therefore, `VehicleRecord` isn't thread safe. The good news is that I don't have to worry as much about starving the connection pool of connection objects.

Formatting

The Result class provides format* methods that allow you to convert query results to different formats like

- CSV
- HTML
- XML
- JSON

So that given a Result of retrieved VehicleModelRecord, I can generate formatted output like so:

```
Result<VehicleModelRecord> vehicleModelRecords =
vehicleManufacturerRecord.fetchChildren(Keys.MANUFACTURER_ID);
vehicleModelRecords.formatJSON();
```

Pretty sweet huh? I should note that this is not a feature for just generated code; you can get the good stuff out of plain SQL jOOQ executions as well. The Record class offers this feature as well, so that you can print whole result sets or individual Records.

Optimistic Locking

To enable jOOQ's optimistic locking mechanism, I configure my DSLContext like so:

```
Settings settings = new
Settings().withExecuteWithOptimisticLocking(true);
DSLContext context = DSL.using(connection, SQLDialect.MYSQL,
settings);
```

Generating code with jOOQ opens up this new, for supporting optimistic locking for transaction management. With "manual" SQL, jOOQ needs to use the SELECT...FOR UPDATE statement to protect concurrent access to rows - this means two trips to the database. With generated code, jOOQ can use the recordVersionFields or recordTimestampFields configuration properties instead:

```
</database>
...
    <recordVersionFields>edens_car\.*\.version
    </recordVersionFields>
...
</database>
Or programmatically
...
.withGenerator(new Generator()
               .withDatabase(new Database()
               .withName("org.jooq.meta.mysql.MySQLDatabase")
               .withRecordVersionFields("edens_car\\.*
               \\.version")
...
```

Using a regular expression, I've stipulated that every table in the edens_car schema that has a version column should be eligible for optimistic locking. It's achievable both programmatically and in XML config. Both of them have the same effect: when two or more transactions are trying to work on the same row, jOOQ will look for the configured column (version) on the affected row. Of course, this means you need to add a version or timestamp column to your table in support of this facility. Whichever transaction has the latest value in recordVersionFields or recordTimestampFields in its copy of that row "wins." The transaction with an older copy of the row will get an org.jooq.exception.DataChangedException when it tries to store, update, or delete its data. If you've worked with Hibernate/Spring Data JPA, you're likely familiar with this mechanism.

Advanced Database Operations

Now that we've had a look at the might of generated code with the jOOQ (lightning and thunder effects!), let's push the envelope a little bit. Now *join* me, as we look at...

Joins

I'm interested in constructing a query that will render a listing of vehicles that gives me a UI representation like so.

Basics

Exterior color	Cadmium Orange
Interior color	Black
Drivetrain	All-wheel Drive
MPG	22–28
Fuel type	Gasoline
Transmission	6-Speed Automatic
Engine	2.0L I4 16V PDI DOHC Turbo

Per vehicle, I want to display

- Vehicle manufacturer name
- Vehicle model
- Model trim
- Current price
- Vehicle color
- Vehicle model year

All this information will need to come from different tables. What does such a query look like?

```
SELECT
        `v`.`vehicle_id` AS `vehicle_id`,
        `v_man`.`manufacturer_name` AS `brand`,
        `v_mod`.`vehicle_model_name` AS `model`,
        `v_trim`.`trim_name` AS `trim`,
        `v`.`vehicle_curr_price` AS `price`,
        `v`.`vehicle_color` AS `color`,
        `v_style`.`vehicle_style_name` AS `style`,
        `v`.`vehicle_model_year` AS `year`,
        `v`.`featured` AS `featured`
FROM  ((((`vehicle` `v`
        JOIN `vehicle_manufacturer` `v_man` ON ((`v`.`vehicle_
        manufacturer` = `v_man`.`manufacturer_id`)))
        JOIN `vehicle_model` `v_mod` ON ((`v_mod`.`vehicle_
        model_id` = `v`.`vehicle_model_id`)))
        JOIN `vehicle_trim` `v_trim` ON ((`v_trim`.`trim_id` =
        `v`.`vehicle_trim`)))
        JOIN `vehicle_style` `v_style` ON ((`v_style`.`vehicle_
        style_id` = `v`.`vehicle_style`)))
```

In this query, I've joined the following tables:

- vehicle
- vehicle_manufacturer
- vehicle_model
- vehicle_trim
- vehicle_style

What does this look like in jOOQ? First, I'm going to compose the columns that I need as a portable bundle:

```
List<Field<?>> fields = Arrays.asList(
                    VEHICLE.VEHICLE_ID,
                    VEHICLE_MANUFACTURER.MANUFACTURER_NAME
                    .as("brand"),
                    VEHICLE_MODEL.VEHICLE_MODEL_
                    NAME.as("model"),
                    VEHICLE_TRIM.TRIM_NAME.as("trim"),
                    VEHICLE.VEHICLE_CURR_PRICE.as("price"),
                    VEHICLE.VEHICLE_COLOR.as("color"),
                    VEHICLE_STYLE.VEHICLE_STYLE_NAME.as("style"),
                    VEHICLE.VEHICLE_MODEL_YEAR.as("year"),
                    VEHICLE.FEATURED);
```

I've put all my desired fields into a neat bundle mostly for the readability advantage. There is a downside here however, because I sacrifice type information by using `Field<?>`. In some scenarios, jOOQ might frown at this because it can't provide the type safety guarantees it would prefer. Note the use of the `as` method to set aliases on the columns. Next comes the actual query. jOOQ offers support for all the major joins, as well as flexibility for how you express the joins. Here, I'm looking to construct an inner join. Here's one way to pull this off:

```
Result<Record> results = context.select(fields).from(VEHICLE)
.join(VEHICLE_MANUFACTURER).on(VEHICLE.VEHICLE_MANUFACTURER.
eq(VEHICLE_MANUFACTURER.MANUFACTURER_ID)) (1)
                    .join(VEHICLE_MODEL).on(VEHICLE.VEHICLE_
                    MODEL_ID.eq(VEHICLE_MODEL.VEHICLE_MODEL_
                    ID))
.join(VEHICLE_TRIM).onKey() (2)
```

```
.join(VEHICLE_STYLE).on(VEHICLE.VEHICLE_STYLE.eq(VEHICLE_STYLE.
VEHICLE_STYLE_ID))
.fetch();
```

So, joining `vehicle`, `vehicle_manufacturer`, `vehicle_model`, and `vehicle_trim` gives me all the information I want. Here's what's new:

1. Starting with my first table, I use the `join` method to connect to the next table. The `on` method helps me define the key on which I want to construct the relation.

2. `onKey` is another variation of `on`. `onKey` is a jOOQ-only enhancement that allows me to skip the part where I have to type in the key column for the relationship. jOOQ will transparently generate an `on` SQL clause by locating a nonambiguous foreign key-primary key relationship between the tables.

    - The `onKey` feature is available only on generated code – jOOQ needs to be sure of the structure of the underlying tables. Your manual SQL poetry won't do, unfortunately.

    - There's a variant available where you can specify the column to use: `onKey(VEHICLE.VEHICLE_TRIM)`.

This yields the standard inner join: a relationship between two or more tables, where the key value is present in all the tables referenced. If you're a stickler for clarity like I am, you may want to explicitly specify that it's an inner join like so:

```
context.select(fields).from(VEHICLE).innerJoin(VEHICLE_
MANUFACTURER).on(VEHICLE.VEHICLE_MANUFACTURER.eq(VEHICLE_
MANUFACTURER.MANUFACTURER_ID))
```

jOOQ supports this join along with

1. Cross-Join with `crossJoin`
2. Left Join with `leftJoin`
3. Right Join with `rightJoin`
4. Full Join with `fullJoin`
5. Partitioned Outer Join `partitionBy` - Oracle only

...and other combinations of joins.

Caution There's a jOOQ bug[8] where non-distinct columns will cause an `InvalidResultException` to be thrown when performing joins with ON or `USING` in some scenarios. One scenario is when the same column name occurs in more than one table in the Join. To get around this, use SQL aliases for columns associated with a Join. Alternatively, you could use the fully qualified (schema.table.column) naming style for your join columns.

Non-SQL Grouping

Consider this scenario: I want to pull the same results as a previous section (vehicle name, model, manufacturer, trim, color, etc.), but grouping the results by manufacturer, so that I can wind up with a `Map` that looks like the following:

```
Map<String, List<VehicleRecord>> vendorMapping
```

[8] `https://github.com/jOOQ/jOOQ/issues/2808`

Here's what the SQL query might look like:

```
SELECT
        `v`.`vehicle_id` AS `vehicle_id`,
        `v_man`.`manufacturer_name` AS `brand`,
        `v_mod`.`vehicle_model_name` AS `model`,
        `v_trim`.`trim_name` AS `trim`,
        `v`.`vehicle_curr_price` AS `price`,
        `v`.`vehicle_color` AS `color`,
        `v_style`.`vehicle_style_name` AS `style`,
        `v`.`vehicle_model_year` AS `year`,
        `v`.`featured` AS `featured`
FROM  ((((`vehicle` `v`
        JOIN `vehicle_manufacturer` `v_man` ON ((`v`.`vehicle_
        manufacturer` = `v_man`.`manufacturer_id`)))
        JOIN `vehicle_model` `v_mod` ON ((`v_mod`.`vehicle_
        model_id` = `v`.`vehicle_model_id`)))
        JOIN `vehicle_trim` `v_trim` ON ((`v_trim`.`trim_id` =
        `v`.`vehicle_trim`)))
<picture confused look caption: huh?>
```

What's that? That's exactly the same query from the section on Joins? There's no GROUP BY statement anywhere in there? Why am I asking you questions when technically I'm talking to myself? Well, friend, this is where the magic of the fetchGroups family of methods comes in.

```
Map<String, Result<Record>> vehiclesGroupedByBrand = context.
select(fields).from(VEHICLE)
.join(VEHICLE_MANUFACTURER).on(VEHICLE.VEHICLE_MANUFACTURER.
eq(VEHICLE_MANUFACTURER.MANUFACTURER_ID))
  .join(VEHICLE_MODEL).on(VEHICLE.VEHICLE_MODEL_ID.eq
  (VEHICLE_MODEL.VEHICLE_MODEL_ID))
.join(VEHICLE_TRIM).onKey()
```

```
.join(VEHICLE_STYLE).on(VEHICLE.VEHICLE_STYLE.eq(VEHICLE_STYLE.
VEHICLE_STYLE_ID))
.fetchGroups(VEHICLE_MANUFACTURER.MANUFACTURER_NAME.
as("brand"));
 vehiclesGroupedByBrand.forEach((manufacturer,vehicles) ->{
        logger.info("Available {} vehicles: \n {}",
        manufacturer,vehicles);
  });
```

I can still construct my join query as normal, but using `fetchGroups`, I can have my results grouped and collected by the alias of a column in the select statement. This is another one of my favorite conveniences with jOOQ. Without this facility, I would have to pick between complex SQL statements or manual composition of the query output to get the same results.

There are many versions of `fetchGroups` that provide superb control over the mapping. For example, I can group the results by POJO:

```
Map<VehicleManufacturer, List<CompleteCarListing>>
vehiclesGroupedByBrand = context
                      .select(VEHICLE_MANUFACTURER.fields())
                      .select(fields) (1)
                      .from(VEHICLE)
.join(VEHICLE_MANUFACTURER).on(VEHICLE.VEHICLE_MANUFACTURER.
eq(VEHICLE_MANUFACTURER.MANUFACTURER_ID))
         .join(VEHICLE_MODEL).on(VEHICLE.VEHICLE_MODEL_
         ID.eq(VEHICLE_MODEL.VEHICLE_MODEL_ID))
.join(VEHICLE_TRIM).onKey()
.join(VEHICLE_STYLE).on(VEHICLE.VEHICLE_STYLE.eq(VEHICLE_STYLE.
VEHICLE_STYLE_ID))
       .fetchGroups(keyRecord -> { (2)
```

```
                return keyRecord.into(VEHICLE_MANUFACTURER).
                into(VehicleManufacturer.class); (a)
              }, valueRecord -> {
                       return valueRecord.into(COMPLETE_CAR_
                       LISTING).into(CompleteCarListing.
                       class); (b)
      });
logger.info("{}",vehiclesGroupedByBrand);
```

It's largely the same join query with some special sauce added:

1. I stack two jOOQ `select` clauses so that the fields in the two statements are available for my purposes later in the query. I call the `fields` method on the `VEHICLE_MANUFACTURER` table to easily load all the fields available in that table. The `fields` variable is the same hand-crafted list of `org.jooq.Field` that I've been using through this section of the book.

    a. The rest of the query contains the same joins as I've been using to demonstrate up till this point.

2. By the time I get to the `fetchGroups` clause, of my jOOQ query, the context has all the fields I need to do the mapping of the query results to POJOs. Note that these are the same POJOs jOOQ generated. This variation of `fetchGroups` allows me to supply instances of `org.jooq.RecordMapper` to construct

    a. The key for the map by correlating the first record into the `VEHICLE_MANUFACTURER` table, and that in turn into the corresponding `VehicleManufacturer` record.

b. The value for the map by correlating the second supplied record into the `COMPLETE_CAR_LISTING` "table" (it's actually a view); that in turn is mapped to the appropriate POJO.

What does the SQL query generated by this jOOQ query look like?

```
select
`vehicle_manufacturer`.`manufacturer_id`,
 `vehicle_manufacturer`.`manufacturer_name`,
.`vehicle_manufacturer`.`status`,
`vehicle_manufacturer`.`version`,
`vehicle`.`vehicle_id`,
`vehicle_manufacturer`.`manufacturer_name` as `brand`,
`vehicle_model`.`vehicle_model_name` as `model`,
`vehicle_trim`.`trim_name` as `trim`,
`vehicle`.`vehicle_curr_price` as `price`,
`vehicle`.`vehicle_color` as `color`,
`vehicle_style`.`vehicle_style_name` as `style`,
`vehicle`.`vehicle_model_year` as `year`,
`vehicle`.`featured`
from `vehicle`
join `vehicle_manufacturer` on `vehicle`.`vehicle_manufacturer`
= `vehicle_manufacturer`.`manufacturer_id`
join `vehicle_model` on `vehicle`.`vehicle_model_id` =
`vehicle_model`.`vehicle_model_id`
join `vehicle_trim` on `vehicle_trim`.`vehicle_manufacturer_id`
= `vehicle_manufacturer`.`manufacturer_id`
join `vehicle_style` on `vehicle`.`vehicle_style` = `vehicle_
style`.`vehicle_style_id`
```

As you can see, stacking my selects simply adds all the available columns to the final SELECT statement. There's also no GROUP BY clause, implicit or otherwise. The grouping is done in memory after the rows have been returned from the database. What does the result look like?

```
VehicleManufacturer (1, Lexus, ACTIVE,
null)=[CompleteCarListing (1, Lexus, ES 350, BASE, 35000.0000,
RED, Car, 2010-01-01, 0), CompleteCarListing (2, Lexus, ES 350,
BASE, 49000.0000, GREY, Car, 2017-01-01, 0)],
VehicleManufacturer (4, Acura, ACTIVE,
null)=[CompleteCarListing (4, Acura, MDX, SPORT, 50000.0000,
BLUE, Car, 2018-01-01, 0), CompleteCarListing (4, Acura, MDX,
BASE, 50000.0000, BLUE, Car, 2018-01-01, 0)
```

Fun fact: fetchGroups, fetchMap, and intoGroup – the three non-SQL grouping functions – will preserve the order of the results as supplied by the query. So if you orderBy, the ordering is preserved across the groupings.

Note Don't forget to generate the equals and hashCode methods on the POJOs (or implement them yourself if they're not generated). The Map data structure needs both hashCode and equals to be able to uniquely identify each of its elements.

When you need to do this kind of grouping in bulk over a large result set, you'll need to revert to the fetchStream. This way, you can take advantage of JDK streaming and parallelization functionality like so:

```
Map<VehicleManufacturer, List<CompleteCarListing>>
vehiclesGroupedByBrand = context
                    .select(VEHICLE_MANUFACTURER.fields())
                    .select(fields) (1)
```

```
.from(VEHICLE)
...
                .join(VEHICLE_STYLE).on(VEHICLE.
                VEHICLE_STYLE.eq(VEHICLE_STYLE.
                VEHICLE_STYLE_ID))
.fetchSize(100) (1)
.fetchLazy() (2)
.collect( (3)
            Collectors.groupingBy( (3a)
                    record -> record.into
                    (VehicleManufacturer.class),
                        (i) (Collectors.
                    mapping(record ->
                    record.into(CompleteCar
                    Listing.class),Collectors.
                    toList())) (ii)
);
```

Breaking out the chain of operation into the `fetchSize` method signals the beginning of lazy business:

1. `fetchSize` sets the maximum number of rows I want the cursor to retrieve in one go. Not setting this might result in the entire result set being loaded into memory.

2. `fetchLazy` officially begins the use of an `org.jooq.Cursor` to efficiently (lazily) stream results.

3. When I go lazy fetching, `fetchGroups` is no longer on the menu. I'll have to take the grouping into my own hands. The `collect` method accepts a `java.util.stream.Collectors`. From this point, you could also go `parallel` because we're in the JDK Streams API territory now.

a. `Collectors.groupingBy`[9] will accept functions that help

   i. Generate the key object.

   ii. Generate the value object and collect the group into a data structure. The handy-dandy `Collectors.toList()` function helps me achieve this.

Altogether, I can now group a large dataset in memory without sacrificing performance. Yes, yes, this goes slightly against the "SQL knows best" mantra,[10] but in the interest of readability, this is a happy medium between complicated window functions and grossly inefficient manual processing.

Batch Operations

I need to insert and/or export one thousand vehicles from the Eden Auto database. What are my options in jOOQ? Well, to start, I can make better use of my database connection:

```
DSLContext context = DSL.using(connection, SQLDialect.MYSQL);
            context.batched(batchedConnectionConfig -> {
               insertVehicle(batchedConnectionConfig);
               insertVehicleModel(batchedConnectionConfig);
               updateVehiclePrice(batchedConnectionConfig):
               //other inserts
            });
```

9 https://docs.oracle.com/en/java/javase/11/docs/api/java.base/java/util/stream/Collectors.html#groupingBy(java.util.function.Function,java.util.function.Supplier,java.util.stream.Collector)

10 Lukas: For a pure SQL approach, consider the MULTISET_AGG function that yields the same results with even better performance: www.jooq.org/doc/latest/manual/sql-building/column-expressions/aggregate-functions/multiset-agg-function/

In the preceding snippet, I've combined multiple dynamically generated insert statements to execute in one shot:

- The `batched` method on `DSLContext` will add the identical statements to a queue. These statements are being teed up for the JDBC batching mechanism to execute in one trip to the database.
- Note that I'm passing the `batchedConnectionConfig` into the query execution methods. The DML methods will need to use this config instead of the original `DSLContext` object.
- They're still executed as independent DML statements, so that each statement yields its own independent `INSERT`, `UPDATE`, or `DELETE`. The advantage comes when jOOQ will delay the execution of these statements for as long as possible before sending them over to the DBMS for execution. This is what's known as the batched connection in the jOOQ API.

Note The batching connection does not kick in when you try to retrieve results from the inserts, for example, generated keys. So, if you have `Settings# returnIdentityOnUpdatableRecord` enabled, calls to `store` on your `UpdatableRecords` will execute immediately instead of waiting for the batch.

I can configure the batch size with the following `Settings` snippet:

```
new Settings().setBatchSize(20);
```

This limits the size of the data sent to my database server in one go: minimizing the risk of overwhelming the network connection or the database itself; you'll need to tune this configuration to match your operational needs.

Explicit Batching

In addition to the batched connection I demonstrated earlier, jOOQ provides convenience batch methods for the operations you'd expect:

- `batch`
- `batchInsert`
- `batchUpdate`
- `batchDelete`
- `batchStore`
- `batchMerge`

With these, I can gain more control over the batching semantics instead of waiting for the `BatchConnection` to do it implicitly. Here's `batchStore` in action:

```
List<VehicleRecord> vehicleRecords = new ArrayList<>();
           //populate list of records to insert
context.batchStore(vehicleRecords).execute();
```

Using any of the other `batchXXX` methods is just as straightforward.[11] You can supply plain SQL, jOOQ DSL statements, or whole entities.

[11] Well, almost. More on this shortly.

There's also the batch mode that yields the same effect:

```
context.batch(
            context.insertInto(VEHICLE, VEHICLE.
            VEHICLE_MANUFACTURER, VEHICLE.VEHICLE_CURR_
            PRICE, VEHICLE.VEHICLE_MODEL_YEAR, VEHICLE.
            VEHICLE_STATUS, VEHICLE.VEHICLE_COLOR,
            VEHICLE.VEHICLE_MODEL_ID, VEHICLE.VEHICLE_
            TRIM, VEHICLE.VEHICLE_STYLE, VEHICLE.
            FEATURED)
                    .values((Long) null, (BigDecimal)
                    null, null, null, null, (Long)
                    null, (Long) null, (Long) null,
                    (Byte) null))
            .bind(4L, BigDecimal.valueOf(46350.00),
            null, "ACTIVE", "BLUE", 13L, 2L, 1L, Byte.
            valueOf("0"))
            .bind(9L, BigDecimal.
            valueOf(83000.00),null, "ACTIVE", "GREY",
            9L, 7L, 1L, Byte.valueOf("0"))
            .bind(9L, BigDecimal.valueOf(77000.00),
            null, "ACTIVE", null, 9L, 7L, 1L, Byte.
            valueOf("0"))
            .execute();
    }
```

The `batch` method allows me to execute my insert statements in bulk by way of value binding. See, instead of separate individual `INSERT` statements, I can use a multi-value insert to execute the batch. The only stipulations are

- To have a stub `values` statement that serves as the "default" values provider. Here, I've used nulls in all available slots.
- I will then use the `bind` method to set up the actual values I want to insert.

Following this, jOOQ will execute the inserts in one shot to the database without the delay that `batchedConnection` uses.

The `batchInsert` and `batchUpdate` methods will generally do what you want them to do as well. They both work with `TableRecord` and `UpdatableRecord`, but there's a catch. The `batchUpdate` method will batch only SQL statements that are identical. So when you have these three `VehicleRecords` being prepped for a `batchInsert`, the results might not be what you expect:

```
VehicleRecord vehicleRecord1 = context.newRecord(VEHICLE);
VehicleRecord vehicleRecord2 = context.newRecord(VEHICLE);
VehicleRecord vehicleRecord3 = context.newRecord(VEHICLE);
vehicleRecord1.setVehicleColor(null);
vehicleRecord2.setVehicleColor("GREY");
vehicleRecord3.setOptions(3L);
context.batchInsert(Arrays.asList(vehicleRecord1,vehicleRecord2,
vehicleRecord3)).execute();
```

There are varying combinations of nulls and actual values for different columns of the same entity `Vehicle` in the preceding snippet. The effect of this is that the resulting SQL from a call to `batchInsert` or `batchStore` will generate functionally separate `INSERT` statements. As a result, jOOQ will **not** batch this update. Instead, it would execute each one individually. In a true batch scenario where you've queued up hundreds or thousands of updates with varying combinations of missing/null fields, you're going to get an unpleasant surprise:

```
<unpleasant surprise image>
```

To be clear, this isn't a jOOQ issue. For a handful of reasons that are outside the scope of this book, most database servers (and JDBC) don't handle null values in `INSERT` and `UPDATE` statements the way you'd expect. To ensure that jOOQ consistently handles my batch inserts and updates the way I'd expect, I will set the `changed` value like so:

```
vehicleRecord1.changed(true);
```

The `changed` flag is an attribute available only with `UpdatableRecord`. It signals to the jOOQ runtime that some fields on this specific entity have been changed. As a result, jOOQ is able to optimize the generated `INSERT` or `UPDATE` statement per batch item.

Tip `UpdatableRecord` provides the previous value of a modified instance. Call the `original` method on the object to get the immediate previous value before a modification.

Batch Import

Yes, you can just inhale or exhale a bunch of data from your database.

Famous batch importer

What does that look like? Let's say I have a CSV that contains rows like the following:

```
vehicle_brand,vehicle_price,model_year,status,color,model_
id,vehicle_trim, style, featured
1,35000.0000,2010-01-01,ACTIVE,RED,1,1,1,0,2021-07-05
13:22:11.0,"","","",""
1,49000.0000,2017-01-01,ACTIVE,GREY,1,1,1,0,2021-07-05
13:22:11.0,"","","",""
1,36000.0000,2018-01-01,ACTIVE,BLUE,1,1,1,0,2021-07-05
13:22:11.0,"","","",""
4,50000.0000,2018-01-01,ACTIVE,BLUE,13,2,1,0,2021-07-05
13:22:11.0,"","","",""
```

How do I get, say, 5000 such CSV rows into my database?

Easy: with the jOOQ Loader API. Observe:

```
context.transaction(txn -> {  (1)
context.loadInto(VEHICLE)   (2)
                    .bulkAfter(50)          (3)
                    .batchAfter(10)        (4)
                    .commitAfter(2)       (5)
                    .loadCSV(csvString) (6)
                    .fields(VEHICLE.fields())
                    .ignoreRows(1)
                    .separator(',')
                    .nullString("")
                    .execute();
      }
);
```

Here's the breakdown:

1. I need to execute the bulk load in a transaction block, which disables `autocommit`.
2. Specify the table that I want to import the CSV data into, using the `Loader` object.
3. `bulkAfter` will configure the size of number of line items contained in inserts sent to the database. Here, I'm asking for each payload to contain 50 `INSERT...VALUES` statements.
    a. `bulkAll` is also an option to send the whole CSV to the database at once. Use with caution and tuning on the database server.

4. `batchAfter` configures the number of individual `INSERT` statements sent at once over the network to the database. Here, I'm asking that ten statements be sent at once. Combined with `bulkAfter`, the configuration could be read as "Add 50 `VALUES` statements to a single `INSERT...VALUES` statement; then send 10 `INSERT...VALUES` statements at once to the server." So in total, a batch execution with these instructions will contain at most 50 rows x 10 statements = 500 rows total.
    a. `batchNone` will execute each `INSERT` individually.
5. `commitAfter` will commit my inserts only after the set number of batches has been reached.
    a. `commitAll` is also an option. On the database server side, all the insert statements sent over the wire will be committed in one large transaction. Be sure your database can support the transaction block size.
6. `loadCSV` tells jOOQ that I want to load the CSV format – JSON is another option.
    a. I define the mapping of the columns in the CSV file to the columns in the database table.
    b. With `ignoreRow` I make jOOQ skip the first row in my CSV, because that's the header row.
    c. The separator symbol for the "columns" in my CSV is set with `separator`.
    d. How I mark a CSV column as null. On encountering this value, jOOQ will substitute `NULL` for any inserts of a "blank" CSV column.

...and then `execute`!

At the time of this writing, CSV and JSON are the only supported file formats for the Loader API. In addition to flat-file formats, I can straight up load my data from memory with the `loadArray` or `loadRecords` methods in the Loader API – loading arrays or jOOQ `Record` respectively. Neato!

Advanced Query Syntax

Even without the gift of jOOQ-generated code and type safety, there are a bunch of powerful and handy SQL features that you can jOOQify. Observe...

Merge and Upsert

How can I conditionally insert or update data in a table depending on whether my insert conflicts with existing data?

Enter the fancy insert twins: **Merge** and **Upsert**. Both help you combine inserts and updates into one SQL statement. No, "upsert" isn't a real word, just a portmanteau of update+insert. The `MERGE` statement is a standard part of SQL, supported by Oracle, SQL Server, DB2, and Sybase, among others. MySQL does not support `MERGE` functionality, but it does provide an alternative. Known as the `INSERT...ON DUPLICATE KEY` statement, it works just like the merge to support

- Inserting rows into a table.
- If the row already exists in the table (and a duplicate key error occurs as a result), the existing record is updated.

In MySQL, my upsert would look like the following:

```
INSERT INTO vehicle_model(vehicle_model_name, vehicle_style_id,
vehicle_man_id)
VALUES('ES 350', 2, 1)
ON DUPLICATE KEY UPDATE vehicle_style_id = 2, vehicle_man_id = 1
```

In jOOQ, I can write the same query thus:

```
context.insertInto(VEHICLE_MODEL,VEHICLE_MODEL.VEHICLE_MODEL_
NAME,VEHICLE_MODEL.VEHICLE_MAN_ID, VEHICLE_MODEL.VEHICLE_STYLE_ID)
                    .values("ES 350",2L, 1L)
                    .onDuplicateKeyUpdate()
                    .set(VEHICLE_MODEL.VEHICLE_STYLE_ID,2L)
                    .set(VEHICLE_MODEL.VEHICLE_MAN_ID,1L)
                    .execute();
```

The onDuplicateKeyUpdate method allows me to define columns to update for any insert attempt that fails due to the data already existing. I still have the option to just straight up ignore any duplicate vehicles for the insertion attempt with onDuplicateKeyIgnore. The SQL equivalent of this directive is the INSERT...IGNORE SQL command,[12] which is exclusive to MySQL. For PostgreSQL, jOOQ supports the newer INSERT...ON CONFLICT statement to achieve the same effects.

So, what if you're not running a MySQL database? Well, you're in luck friend – MERGE is going to save you. jOOQ will transparently translate any usage of onDuplicateKeyUpdate and onDuplicateKeyIgnore to a MERGE statement, where the backing database is not MySQL:

[12] With some caveats: https://github.com/jOOQ/jOOQ/issues/5211

```
context.mergeInto(VEHICLE_MODEL)        (1)
                        .using(selectOne())                    (2)
                        .on(VEHICLE_MODEL.VEHICLE_MODEL_NAME.
                        eq("ES 350"))       (3)
                        .whenMatchedThenUpdate()       (4)
                        .set(VEHICLE_MODEL.VEHICLE_STYLE_ID,3L)
                        .set(VEHICLE_MODEL.VEHICLE_MAN_ID,1L)
                              .whenNotMatchedThenInsert(VEHICL
                        E_MODEL.VEHICLE_MODEL_NAME,VEHICLE_MODEL.
                        VEHICLE_MAN_ID,VEHICLE_MODEL.VEHICLE_STYLE_
                        ID)      (5)
                        .values("ES 350",2L, 1L)
                        .execute();
```

It's a one-to-one translation where

1. It starts with the `mergeInto` node in the fluent chain.

2. The standard `MERGE` syntax requires source and destination tables for the data. In this snippet, my data is manually built and not coming from another database table. Therefore, I won't have a source table to provide. That's why I'm using the `selectOne()` method on `DSLContext`. This is a convenience feature (one of many) that generates a query from a pseudo table (like Oracle's `DUAL` table). For any other scenario, you would supply an actual table in this position.

    a. You can use `Records` and subqueries in the `using` clause as well. They just need to be converted to tables. The `table` function will convert almost anything into a table for the purposes of a jOOQ query, for example:

```
List<VehicleModelRecord> vehicleModelRecords = ...
...
            context.mergeInto(VEHICLE_MODEL)
                      .using(table(vehicleModelRecords))
                      ...
```

3. I then define the condition that I want to use to determine whether a row is a duplicate or not.
4. That being set, `whenMatchedThenUpdate` does what it says: rows that match the condition will be updated with the following `set` data.
5. `whenNotMatchedThenInsert` will kick in if no matches are found; a new row will be inserted.

Tip `onDuplicateKeyIgnore` and `onDuplicateKeyUpdate` are all available for the **Loader** API as well. `onDuplicateKeyError` is available for the Loader API only. This means that you can apply even more flexible upsert semantics to bulk loading of data.

`MERGE` is definitely more powerful than the `INSERT...ON DUPLICATE KEY` or `INSERT...IGNORE` statements. For example, in some database dialects (Oracle, DB2, Sybase), you can `DELETE` rows that fail the matching condition.

Window Functions

I'd previously used window functions in the previous chapter to construct a Common Table Expression (CTE) to calculate the median car price in the inventory. I didn't go into any detail, so here we are again.

While this isn't a SQL textbook,[13] window functions are an enigmatic and enormously powerful toolkit that warrant some explanation. Here's a look at the vehicles in my inventory:

vehicle_id	brand	model	trim	price	color	style	year	featured
1	Lexus	ES 350	BASE	35000.0000	RED	Car	2010	0
2	Lexus	ES 350	BASE	49000.0000	GREY	Car	2017	0
3	Lexus	ES 350	ES 350 ASE	36000.0000	BLUE	Car	2018	0
4	Acura	MDX	SPORT	50000.0000	BLUE	Car	2018	0
5	Acura	MDX	SPORT	55020.0000	WHITE	Car	2017	0
6	Acura	MDX	SPORT	45000.0000	WHITE	Car	2013	0
7	Toyota	Corolla	CE	18550.0000	BLUE	Car	2020	1
8	Toyota	Corolla	CE	24000.0000	RED	Car	2019	0
9	Toyota	Camry	XLE	30000.0000	BLUE	Car	2021	1
10	Acura	MDX	SPORT	46350.0000	BLUE	Car	2021	1
11	Mercedes-Benz	S500	BASE	83000.0000	GREY	Car	2021	1
12	Mercedes-Benz	S500	BASE	77000.0000	WHITE	Car	2016	1
13	Acura	MDX	SPORT	46350.0000	BLUE	Car	2014	0
16	Acura	MDX	SPORT	46350.0000	BLUE	Car	2019	0

Nothing special, just a `SELECT *` of the vehicles from the `complete_car_listing` table. I'm interested in getting a report that looks a little something like this from that table:

vehicle_id	brand	model	year	price	price_rank
5	Acura	MDX	2017	55020.0000	1
4	Acura	MDX	2018	50000.0000	2
10	Acura	MDX	20 2021	[illegible]50.0000	3
13	Acura	MDX	2014	46350.0000	3
16	Acura	MDX	2019	46350.0000	3
6	Acura	MDX	2013	45000.0000	4
19	Acura	MDX	2015	42450.0000	5
2	Lexus	ES 350	2017	49000.0000	1
3	Lexus	ES 350	2018	36000.0000	2
1	Lexus	ES 350	2010	35000.0000	3
11	Mercedes-Benz	S500	2021	83000.0000	1
12	Mercedes-Benz	S500	2016	77000.0000	2
9	Toyota	Camry	2021	30000.0000	1
8	Toyota	Corolla	2019	24000.0000	2
7	Toyota	Corolla	2020	18550.0000	3

[13] Editor's note: That could come later. ☺

In the screenshot above, I have a report that shows me the individual vehicles in the inventory and their prices, among other things. Here are the key fields:

- `vehicle_id` obviously refers to a specific vehicle. I also have some individual vehicle details like `model`, `year`, and `price`.
- The `price` column is the individual vehicle's price
- Then I have a `price_rank` column that shows me how the individual vehicle `price` ranks against the prices of similar vehicles of the same `brand`.

A regular group function like AVG or MAX will collapse all the data into a single value like "here's the max price of any Toyota in the inventory." Using window functions, I can say "here are the prices for the individual vehicles in the inventory, but for each row, I want to display the rank of the vehicle's price."

TL;DR: Window functions let you combine the summarization capabilities of group functions while retaining the ability to display the individual rows that make up the group values.

It's...not the easiest thing to explain. So, let me show you how a window function can yield the results I'm interested in. Here's what the SQL query looks like:

```
select vehicle_id,brand,model, year, price, avg(price) as
avg_price, dense_rank() over (partition by brand order by
avg(price) desc) as price_rank
from edens_car.complete_car_listing
group by brand, model,vehicle_id
```

1. The preceding query has standard SQL components and clauses - the AVG group function, the GROUP BY clause, etc.

2. The DENSE_RANK function is what gives an ordinal number to the rows returned from the result. It's one in a family of ranking functions; there's also RANK and ROW_NUM functions that provide similar features but with some differences in how they handle ties between rows.

3. OVER is what marks the start of a window function. It stipulates the range over which the window function needs to be applied. In this case,

    a. The ranking needs to happen based on the prices of the vehicles.

    b. Optionally, I want the rankings to be further grouped by brand. This way, the rankings are within a manufacturer's vehicle range. Instead of saying "show me the ranking of all the vehicle prices," I'm saying here with PARTITION, "group the rankings into buckets per vehicle manufacturer."

All told, I can see

1. Individual car records and their details

2. The result of grouping car records together by the average of their prices

3. The rank of an individual vehicle's price relative to the average of its group

A moment to catch our breath...and then we look at what this would look like in jOOQ:

```
context.select(COMPLETE_CAR_LISTING.VEHICLE_ID,
                                   COMPLETE_CAR_LISTING.BRAND,
                                   COMPLETE_CAR_LISTING.MODEL,
                                   COMPLETE_CAR_LISTING.YEAR,
```

```
                                avg(COMPLETE_CAR_LISTING.PRICE).
                                as("avg_price"),
                                rank().over(partitionBy(COMPLETE_
                                CAR_LISTING.BRAND)
                                        .orderBy(avg(COMPLETE_CAR_
                                        LISTING.PRICE).asc()))
                                        .as("price_rank")
                        )
                        .from(COMPLETE_CAR_LISTING)
                        .groupBy(COMPLETE_CAR_LISTING.BRAND,
COMPLETE_CAR_LISTING.MODEL, COMPLETE_CAR_LISTING.VEHICLE_ID);
```

I know, it seems like a lot of code, but you can read it pretty much as a one-to-one mapping with the SQL version of the query. The `rank` function and everything that follows it are supplied by the one and only `DSL` class.

Phew! Here's a cute duckling for sticking with the book thus far.

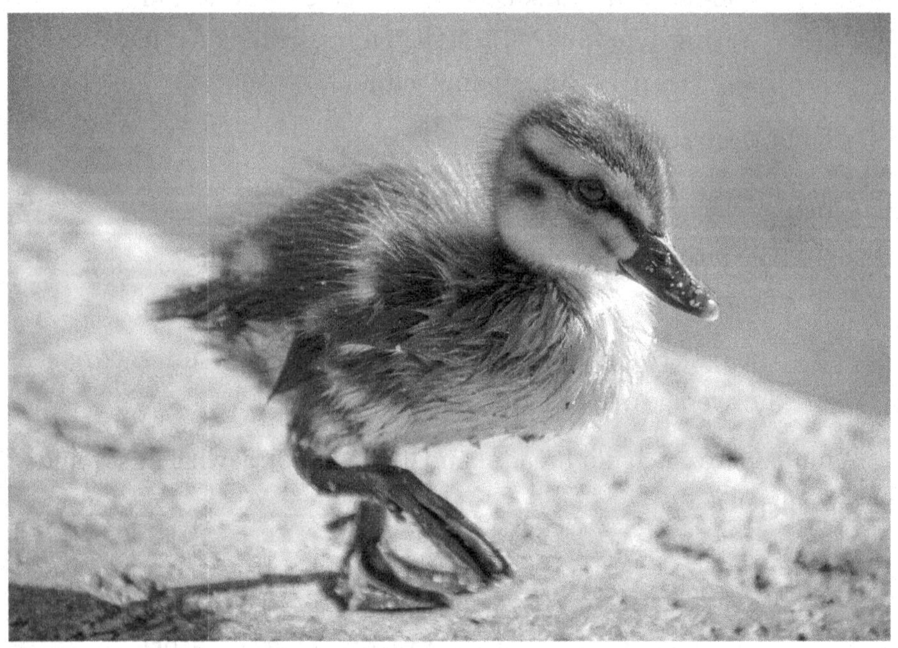

Rubber Duck Debugging: The live action movie!

CHAPTER 4

Integrating with jOOQ

Also known as "eating your cake and having it too." If you're starting afresh with jOOQ in a new project, congratulations and *bonne aventure*! And if you have an existing project with other technologies and APIs? jOOQ could still be of mighty service. To recap, jOOQ does the following specific things exceedingly well:

- **Generating highly expressive and typesafe, reusable SQL** so that you can rest assured that your SQL is always correct.
- **Generating Java classes from database entities** so that you never have to manually construct another entity, DTO, or Active Record again.
- **Smoothly managing database dialects, quirks, and shortcomings** so that you don't have to get bothered by the differences between different database vendors. Superlative portability!

...and of course, staying aligned with the performance and scalability of interests of your database server. That being said, jOOQ isn't looking to be your one and only love. I mean, it would be nice, but if you already have certain technologies in use, jOOQ is more than happy to share responsibilities.

T. Koleoso, *Beginning jOOQ*, https://doi.org/10.1007/978-1-4842-7431-6_4

*As long as we all understand who's the **real** friend of the database in this alliance*

In this chapter, we're going to take a look at how jOOQ can supercharge your application implementations by lending its unique capabilities to existing APIs and libraries.

Java Persistence API with jOOQ

Java Persistence API (JPA) is the JakartaEE specification that defines how database objects can be mapped to Java, a.k.a. Object-Relational Mapping (ORM). It lays out how implementation APIs should handle translating database components, SQL, and other database stuff into Java classes, interfaces, and in reverse. It defines the expected behavior of reference implementations during specific circumstances. It also defines the Java

Persistence Query Language (JPQL), a SQL-flavored query syntax that attempts to replicate the idioms of SQL, but for Java classes. We then look to vendors like

- Redhat/Hibernate
- Eclipse/EclipseLink
- Oracle/TopLink
- OpenJPA

Ultimately, the expectation is that industry vendors will implement a functional API following the guidance of the specification. Keyword there is "guidance" - the specification is a guideline, and vendors can and typically do break rules of the spec. Alternatively, they could implement some specification-defined functionality, but in a non-standard way. What this means is that your mileage may vary from JPA implementation to implementation.

But you probably already knew all this.

We're not going to explore JPA in any great detail here; just want to answer one question: what can jOOQ do for you in a JPA world?

Generate JPA Entities

The fundamental unit of work in JPA is the entity. The JPA entity is a Java class that you use to represent a database table or rows from that table. Because this isn't a JPA textbook, I'm not going to go into too much detail about JPA entities. Suffice it to say, jOOQ can create some basic JPA entities for you. All you need to do is ask:

```
<generate>
            <jpaAnnotations>true</jpaAnnotations>
</generate>
```

It's really that simple. Flag `jpaAnnotations` "on" in your jooq code generator settings and you'll get something like the following:

```
/**
 * This class is generated by jOOQ.
 */
@Entity
@Table(
    name = "vehicle",
    schema = "edens_car",
    indexes = {
        @Index(name = "veh_manufacturer_id_idx", columnList =
        "vehicle_manufacturer ASC"),
        @Index(name = "veh_model_id_idx", columnList =
        "vehicle_model_id ASC"),
        @Index(name = "veh_style_idx", columnList = "vehicle_
        style ASC")
    }
)
public class Vehicle implements Serializable {
    private static final long serialVersionUID = 1L;
    private Long          vehicleId;
    private Long          vehicleManufacturer;
    private BigDecimal    vehicleCurrPrice;
    private LocalDate     vehicleModelYear;
    private String        vehicleStatus;
    private String        vehicleColor;
    private Long          vehicleModelId;
    private Long          vehicleTrim;
    private Long          vehicleStyle;
    //more fields
```

```
    public Vehicle() {}
    //getters, setters, constructors, toString etc
}
```

The most important annotation here is `@Entity`. This is what signifies to the JPA runtime that instances of this class should be managed by the JPA runtime. This has significant implications for the way instances of this class are regarded by the JPA runtime. From the moment an instance of this entity class `Vehicle` exists, the JPA runtime is paying attention. Any changes on the entity, any new instances of it, retrievals from the DB, etc. all tracked by the JPA runtime. When two threads are trying to modify the underlying table row that backs a specific instance of `Vehicle`, it's the JPA runtime's job to make sure only one or none of them succeed in making changes.

And now, a word from our sponsors.

No, not you sir!

I'm a strong advocate of treating instances of your JPA entities the same way you would treat the underlying database row data. It's a very pervasive but insidious code smell to handle these entities the way you would handle a "dumb" object like a POJO or a Data Transfer Object (DTO). See, because entities are live, managed objects, you run the risk of

1. Accidentally changing the state of the object in the normal course of process execution.

2. Inducing a state management exception like `StaleObjectStateException` if you hold on to an instance of the entity for too long in a read-only operation. This is especially likely to happen in distributed environments and microservices. One thread simply wants to read some data, maybe send it as a web service response. Another thread concurrently wants to make changes to the underlying data for the same entity. One of those threads is going to have a bad time.

3. Leaking data when you use the same entity class for database operations, as well as web service responses, or persisting it to a different format like JSON. You're going to indiscriminately transmit table columns in several directions.

TL;DR: Separate the concerns between your POJO needs and your ORM needs. They are not the same type of class.

One workaround is to run the code generator twice: once with `jpaAnnotations` set to `false` and another with it set to `true`. Remember to change the output packages between both runs.

In addition to vanilla JPA annotations, jOOQ can add

- The `Serializable` interface with `<serializablePojos>true</serializablePojos>`
- JPA support for a specific version with `<jpaVersion>2.2</jpaVersion>`

Neato.

Generate *from* JPA Entities

Yes, you read that right: jOOQ can get you started with jOOQing, even if you don't have an actual database. If you have JPA entities but no database, jOOQ can still generate code for ya. Considering my rant against reusing entities as DTOs or POJOs, this comes in super handy. This way, your JPA entities could be already pre-generated and packaged maybe as a JAR; all you'd need to do is generate POJOs from those entities and you'd be on easy street. Observe.

Start by adding the following Maven (or equivalent Gradle, etc.) entries to your project:

```
<dependency>
        <groupId>org.jooq.pro-java-11</groupId>
        <artifactId>jooq-meta-extensions-hibernate</artifactId>
        <version>3.15.1</version>
</dependency>
```

That'll pull in the jooq hibernate extensions kit. Next up, some configuration changes to the code generator itself:

```
<database>
      <name>org.jooq.meta.extensions.jpa.JPADatabase</name> (1)
    <properties>
```

```
    <!-- A comma separated list of Java packages, that
    contain your entities -->
        <property>
            <key>packages</key>
            <value>com.apress.samples.jooq.jpa.entity, com.
            apress.samples.jooq.ext.jpa</value> (2)
        </property>
<!-- The default schema for unqualified objects:
- public: all unqualified objects are located in the PUBLIC
(upper case) schema
- none: all unqualified objects are located in the default
schema (default)
This configuration can be overridden with the schema mapping
feature -->
        <property>
            <key>unqualifiedSchema</key>  (3)
            <value>none</value>
        </property>
    </properties>
</database>
```

What's all this then?

1. For generating from JPA entity classes, a name change is necessary. `org.jooq.meta.extensions.jpa.JPADatabase` defines the source of data for the generator. Contrast this with the `org.jooq.meta.mysql.MySQLDatabase` I've been using up till now, because my code was being generated from an actual database.

2. I specify the packages that jOOQ should scan to be able to parse JPA entity classes.

3. How should jOOQ handle entities that don't have schema data? The `unqualifiedSchema` property accepts `none`, meaning all entities missing schema info will be put in the default schema. `public` is also valid, meaning these entities will be put in the public schema by default. You could still override all this with `SchemaMapping` as well.

For my crusade against abusing entity classes as DTOs, this is another good workaround especially when you already have legacy JPA entity classes. Simply generate POJOs from your existing JPA entities and skip the `jpaAnnotations` directive in your code generator configuration.

Generate SQL Queries

Of course, the most obvious use case. jOOQ will always out-SQL anything JPA could conjure up. So, it stands to reason that when you want to take your database seriously, you should consider delegating SQL query generation to jOOQ. JPA provides a number of opportunities to supply your own SQL. Consider our beloved vehicle select query:

```
Query jooqQuery = DSL.using(SQLDialect.MYSQL,new Settings()
                .withRenderQuotedNames(RenderQuotedNames.NEVER))
                .select(VEHICLE.VEHICLE_ID, VEHICLE.VEHICLE_
                COLOR, VEHICLE.VEHICLE_CURR_PRICE)
                .from(VEHICLE)
                .where(VEHICLE.VEHICLE_MANUFACTURER.
                eq(param("vehicle_manufacturer", Long.class))).
                getQuery();
```

The preceding jOOQ statement

- Uses the `DSL` class to configure the dialect of the soon to be generated SQL statement. It also specifies that quotation marks shouldn't be used in the generated SQL - this could become important, depending on the dialect that's configured for your JPA implementation. The regular double quotes (") could cause Hibernate to choke, for example.
- Selects some fields from the `vehicle` table, but instead of executing it, I'm obtaining an instance of `org.jooq.Query`. This is the parent interface of the jOOQ representation of all of the SQL statements. Did you notice how I'm not using the `DSLContext` here? Instead, I'm using the `DSL` class directly to create my select statement. This means that I don't need to go about constructing a `DSLContext` or a JDBC connection just so I can build a jOOQ SQL query.
- Binds the `VEHICLE.MANUFACTURER` column as a query parameter with the `param` function. What this means is that I can supply a dynamic value at runtime.
- The `getQuery` method at the end yields a `Query` object from which I can then obtain the plaintext SQL statement, among other things.

How can this help within the JPA world?

JPA has a set of opportunities for you to supply your own SQL query. Why would you want to do this? Well, the simple fact of the matter is that for anything more complex than a straightforward `SELECT` statement from a couple of tables, JPA isn't the best option, especially at scale. If you need to use Common Table Expressions, inline views, window functions, etc., you're going to need to craft your own SQL. Hierarchical queries are not on the JPA menu. As much as JPQL can try, it supports only a subset of the SQL specification. This is where your Query object comes in.

```
@PersistenceContext
 EntityManager entityManager;  (1)
...
javax.persistence.Query nativeQuery = entityManager.
createNativeQuery(jooqQuery.getSQL());  (2)
int parameterCount = 1; //JDBC parameter values begin their
index at 1, not 0.
 long vehicleStyle = 4;
       for(Parameter parameter: nativeQuery.getParameters()){
            nativeQuery.setParameter(parameterCount++,
            vehicleStyle); (3)
       }
List<Vehicle> resultList = nativeQuery.getResultList(); (4)
logger.info("Results: count: {} \n list: {}",resultList.
size(),resultList.toArray());
```

Alrighty then; let's dig into this:

1. The `EntityManager` is the gateway into the JPA runtime, also known as the `PersistenceContext`. Did I mention how much I live for the Context Object pattern? This is one of those. All the database rows mapped from the database can be obtained from this object. Pretty much anything you want to do with JPA can be initiated from this. There are multiple ways to obtain an instance of this object, depending on what platform you're working with (JakartaEE, Spring Data, Quarkus, etc.); I won't be going into that detail here.

2. The `EntityManager` object provides the `createNativeQuery` method that allows me to supply custom SQL. This is where my `org.jooq.Query` object comes to shine. I use the `getSQL` method to obtain plaintext SQL generated from my jOOQ query.

3. Because I've defined a query parameter on the jOOQ query, the JPA query automatically inherits the parameter via the plain String SQL I passed to it. This means that I can dynamically set values for each available `Parameter` as recognized by the JPA `Query` object. This is a particularly flexible operation overall, for example: I can refer to my query parameters by name, in addition to the index value.

4. Finally, I can execute the SQL statement and retrieve my query results with the `Query#getResultList`. This method can return a list of JPA entity classes or a list of `Objects` that I can transparently cast to any class of my choosing. Here, I've chosen to use the `Vehicle` POJO class that jOOQ generated for me. This is a non-attached, unmanaged java object, so I don't have to worry about accidentally modifying underlying database data via the results of this query.

There are other opportunities to take advantage of reliably sourced, certified conflict-free and gluten-free SQL in the JPA world. You can pass in a JPA entity class to the `createNativeQueryMethod`.

```
Query nativeQuery = entityManager.createNativeQuery(jooqQuery.
getSQL(), VehicleEntity.class)
```

With this approach, any instances of `VehicleEntity` that are returned by `EntityManager` are managed objects - if you make changes to the state of those objects, you will affect the data in the underlying db row. This will map the query results to instances of that class, provided the column names and other things match.

What about when the database columns don't line up with your class declarations? Maybe you're using a column alias, or you want to return multiple entity types from one statement? What about when you don't want to use a JPA entity class at all? I for one definitely enjoy not worrying about accidentally modifying table data via an entity. I want disconnected objects for read-only purposes.

Behold, the three JPA horsemen of SQL result mapping!

"lol what's a SQL?"

Okay, seriously, it's these three annotations:

1. `@SqlResultSetMapping` defines the existence of a need to map results from a SQL query into a java object. This annotation can be applied to any JPA class with the `Entity` annotation. After defining this annotation on a class, you can refer to it by name. Stick with me to see this in action.
2. `@ConstructorResult` was introduced with JPA 2.1 so that we can use JPA to construct unmanaged Java objects/entities. Prior to this, everything had to be a managed JPA entity. See my previous caveat for why this can become a bad thing. With this annotation, even when you provide a class annotated with `@Entity`, the JPA runtime will ignore it and not attempt to manage anything about the results of this construction.
3. `@ColumnMapping` allows you to map the columns in a SQL query result to the fields of a non-JPA entity, that is, a POJO. This is how you define the mapping of column aliases and other non-conforming names from SQL results to the fields of a Java class. In JPA lingo, such columns are called **scalar columns**.

So, how do these work together? Check this out: given that I've run my jOOQ generator and obtained a JPA-annotated POJO `com.apress.jooq.generated.tables.pojos.VehicleModel`, I can put the mapping annotations to work like so:

```
@SqlResultSetMapping(name="nonJPAManagedVehicleModel", (1)
        classes = {
        @ConstructorResult(targetClass =  com.apress.jooq.
        generated.tables.pojos.VehicleModel.class,    (2)
```

```
        columns = {                      (3)
                        @ColumnResult(name="vehicle_
                        model_id"),
                        @ColumnResult(name="vehicle_
                        model_name"),
                        @ColumnResult(name="vehicle_
                        style_id"),
                        @ColumnResult(name="vehicle_
                        man_id"),
                        @ColumnResult(name="version")
                         })
        })
@Entity
public class VehicleModel implements Serializable {
...
}
```

Okay, buckle up while I explain what's going on here:

1. It all starts with `@SqlResultSetMapping` here, where I say: "I want to define a custom mapping between a SQL statement and a POJO. I've **named** the query `nonJPAManagedVehicleModel` because that's how I roll."

2. I then define the classes that are involved in this custom mapping. For this example, I'm interested only in the POJO `VehicleModel`. Here's where things get a little verbose.

    a. I need to describe a suitable constructor for the JPA runtime to be able to create instances of my POJO class with `@ConstructorResult`. Given this mapping, JPA knows what to do with the results from the query.

   b. Remember: even though this class is technically a JPA entity class, when I use it in this context, JPA will not treat the results of this query as managed entities, which is pretty rad in my opinion.

3. `@ColumnResult` helps me map the name that is present in the query results to fields in the `VehicleModel` POJO class. How does JPA know which field on the class to map the column to? By the **position** of the column in the list of `columns`. The JPA runtime will look for a suitable constructor matching the description here and just pick the named columns to be passed into the constructor.

At the end of it all, I can then use my named query like so:

```
entityManager.createNativeQuery(vehicleModelQuery.getSQL(),
"nonJPAManagedVehicleModel");
```

The JPA runtime will attempt to look my SQL mapping using the name I've supplied. That gives it all the information it needs to execute the query and build a list of result objects.

This is fairly verbose, so don't worry if it doesn't sink in all at once - go over it as many times as you can. The simpler description of what's going on is to tell JPA

1. What POJO or entity classes you want to map to
2. Which column names should be used from the SQL result set
3. Which constructor to use on your POJO class
4. Which columns should be used in the constructor of the POJO class

All told, these facilities allow you to package your SQL queries in very portable and flexible deployment units; take database dialects into account as well as guarantee the effectiveness of your queries.

Caution As at the time of this writing, jOOQ has a bug[1] that makes it require the `@Column` annotation on JPA entities. This won't be an issue with jOOQ-generated entity classes; but if you're bringing your own JPA entities to the jOOQ party, be sure to add `@Column` to the fields on that entity. Some weird stuff happens otherwise (e.g., column values not being mapped to result objects).

Now, we say goodbye to the horsemen of result set mapping.

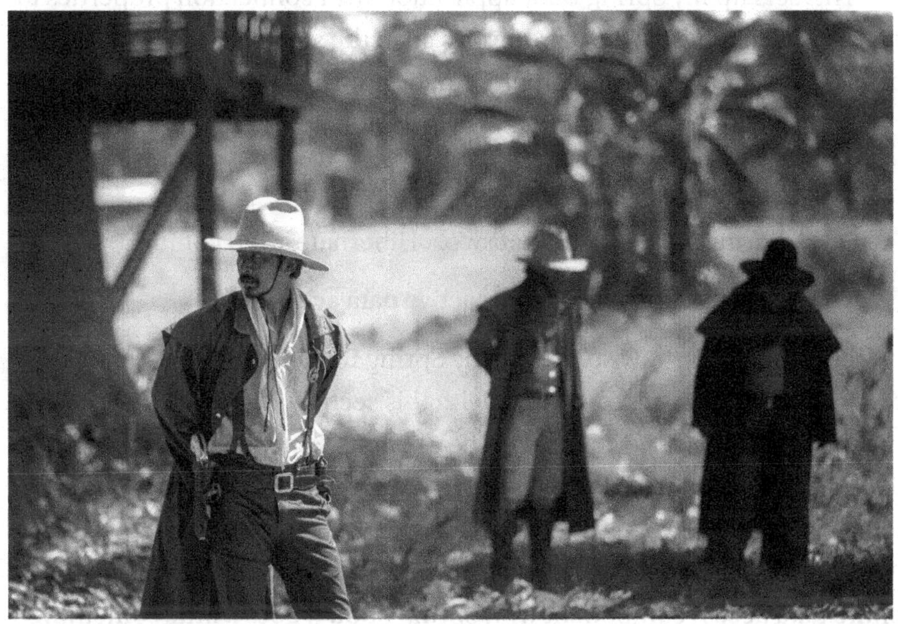

Aww! Chin up, lads!

[1] https://github.com/jOOQ/jOOQ/issues/4586

Spring Boot and jOOQ

Spring Boot is the current king of the enterprise Java development hill. *Current.* There's almost nothing you can't do with the Spring platform and I'm not even going to attempt broaching its many features here. Let's see how jOOQ can spruce up your Spring Boot application. But first, some configuration:

```
spring.datasource.driver-class-name=com.mysql.cj.jdbc.Driver
spring.datasource.url=jdbc:mysql://localhost/edens_car
spring.datasource.username=username
spring.datasource.password=thisisaterriblepassword
spring.jpa.show-sql=true
```

This sets up my Spring Boot application with connection properties to my MySQL database. These properties go in the standard `application.properties` file. There's the programmatic equivalent as well.

Spring supports SQL data access via the following components:

- Spring Data JDBC for vanilla JDBC access
- Spring Data JPA for that sweet Hibernate + JPA combo
- Spring Data R2DBC for reactive data access

Because of just how sprawling the Spring Boot platform is, I'm going to try to keep things nice and tight with this section. Also note that pretty much everything we've covered in the previous section, as well as this one, all apply to Hibernate. Hibernate is the JPA implementation that powers a lot of Spring framework's data access powers.

First thing to know is that you can wholesale install jOOQ as your entire data access component with Spring Boot. It goes a little something like this.

Configure jOOQ in Spring Boot

Let's make jOOQ available everywhere in the Spring application context with a programmatic configuration setup:

```
@Configuration                                                        (1)
@EnableTransactionManagement
public class JdbcConfig extends AbstractJdbcConfiguration {
    @Autowired
    private DataSource dataSource;                          (2)

    @Bean
    DataSourceConnectionProvider connectionProvider() {
        return new DataSourceConnectionProvider(new
        TransactionAwareDataSourceProxy(dataSource));
    }

    @Bean
    DSLContext dsl() {
        return new DefaultDSLContext(dslConfig());                (3)
    }

    private org.jooq.Configuration dslConfig() {
        DefaultConfiguration defaultConfiguration = new
        DefaultConfiguration();
        defaultConfiguration.set(dataSource)
                .set(SQLDialect.MYSQL)
                .set(DefaultExecuteListenerProvider.
                providers(new QueryRuntimeListener()));
        return defaultConfiguration;
    }
}
```

There's a fair bit of Spring framework boilerplate going on here, but I'll be focusing on the jOOQ-relevant bits:

1. I set up my Spring configuration bean with the `@Configuration` annotation and other standard Spring framework componentry like `@EnableTransactionManagement` to let Spring manage my database transactions; `AbstractJdbcConfiguration` so my config class can inherit even more boilerplate. It's boilerplate smorgasbord.

2. I use Spring's dependency injection to obtain a `DataSource` instance. `DataSource` is a more mature, scalable, and robust representation of my database connection and pool, managed by the Spring Boot runtime. This will be supplied here because I've already configured my database properties in the standard `application.properties` file.

3. I define a method that can construct an instance of `DSLContext` on demand. Adding the `@Bean` annotation marks this as a factory method for Spring Boot. This means I can obtain a fresh instance of `DSLContext` anywhere inside my Spring application.

With this setup, I can obtain a `DSLContext` anywhere in the application:

```
@Autowired
DSLContext context;

public void selectWithJooq(){
    context.selectOne();
}
```

And then I can jOOQ away to my heart's content. The preceding snippet can be extended to produce a new DSLContext instance per request, support multi-tenancy, and so much more. If you can dream it, jOOQ could probably make a best effort attempt at making it happen. Not to mention the DAOs that jOOQ can generate for you. Nice.

How about custom SQL queries?

With Custom SQL

If there's custom SQL that needs writin', jOOQ's gonna be a-generatin'. To use custom queries with Spring Data JPA, I start by creating a Repository:

```
public interface VehicleModelJooqRepository extends
CrudRepository<VehicleModel, Long> { (1)
    @Query(nativeQuery = true, name="CustomDynamicSQL")
    List<VehicleModel> findVehicleModelByVehicleManId(long id);
}
```

Allow me to explain:

1. I extend CrudRepository as part of the contract for using Spring Data JPA's repository feature. Specifying Vehicle and Long as the types for this interface, I'm informing the Spring Data runtime that this interface will be used to retrieve VehicleModel from the vehicle_model table.

2. I define a findVehicleModelByVehicleManId that accepts a long parameter corresponding to the vehicle_man_id to filter results by

   a. Crucially, I'm using the @org.springframework.data.jpa.repository.Query annotation. Spring Data JPA allows me to specify a plaintext SQL query in this annotation; alternatively, I can define the query elsewhere, and with some Spring magic, it'll be picked up. Stay tuned to see how. So far, this JPA repository is expecting to find a native query named "CustomDynamicSQL" somewhere in the PersistenceContext.

b. Each argument I pass to the query method findVehicleModelByVehicleManId will in turn be fed as query parameters to the native query that this method will execute. This is important because you either have to match the positions of the method arguments to the positions of the query parameter in the plain SQL; alternatively, you can use the @Param annotation to name-match your parameters against their SQL equivalent.

Now, I need to wire up my custom SQL query, sponsored by jOOQ. To actually plug my SQL query into the JPA runtime, I turn back to our old friend, EntityManager:

```
javax.persistence.Query nativeQuery = entityManager.
createNativeQuery(jooqQuery.getSQL());
entityManager.getEntityManagerFactory().addNamedQuery("CustomDy
namicSQL",nativeQuery);
```

Having obtained an instance of javax.persistence.Query from my org.jooq.Query:

1. I obtain an EntityManagerFactory from the EntityManager.
2. The addNamedQuery method was added to JPA 2.1 to allow dynamic construction of named queries. With this, I need to supply

a. A name for the query by which Spring Data JPA can look it up. Note how I'm using the fully qualified name of the method I defined in my `Repository` interface from earlier. This is how Spring Data JPA will attempt to look up the named query based on the `@Query` annotation I added to my custom `Repository` method.

b. The actual SQL query to be executed.

Spring Boot takes care of the rest from here. I can just inject my custom repository and use it as I want:

```
@Autowired
VehicleModelJooqRepository vehicleModelRepository;
...
List<VehicleModel> modelByVehicleManId =
vehicleModelRepository.getVehicleModelByVehicleManId(vehicleMan
ufacturer);
```

...aaand that's it! This dynamic SQL feature is in addition to the standard JPA features we've already explored - Spring supports those too.

jOOQ Spring Boot Starter

Spring Boot supposedly offers a starter that helps you bootstrap your Boot project with jOOQ.

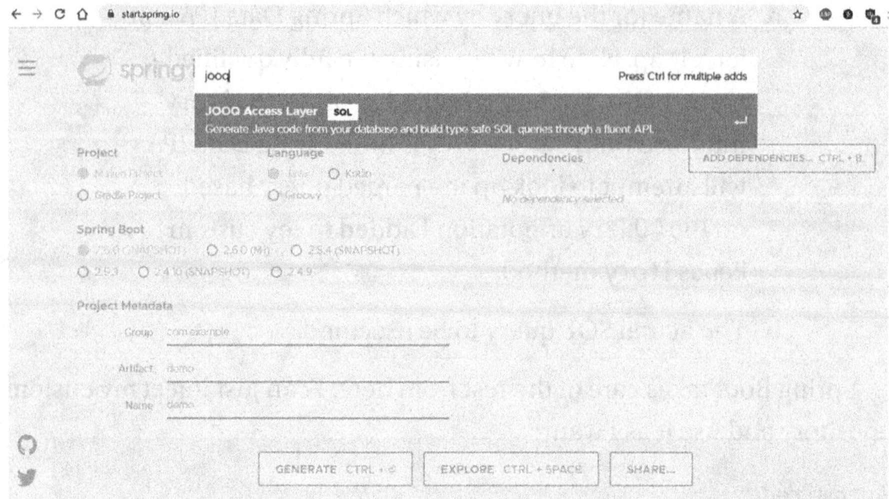

start.spring.io

In practice, I tried to bootstrap with this, including the Spring Data JPA and JDBC modules. It doesn't look to be very effective to me because

- The starter doesn't (currently) include any jOOQ dependencies.
- The code stubs it generates don't even include any references to jOOQ (see the previous discussion).

So, maybe hold off on this one for a bit.

Quarkus and jOOQ

Quarkus[2] is *the* premier cloud-native, container, and Kubernetes-first microservices platform. It supports pretty much anything you'd want to do with a Java web service platform. You can integrate your existing JakartaEE or Spring beans and use the same programming components to get

[2]`https://quarkus.io`

- Blazing fast startup times
- Low memory footprint
- Tight integration with features and components of major cloud providers like AWS, Google Cloud, and Azure
- Lightweight deployment packages
- Reactive-first programming style
- Kotlin and Scala compatibility

Quarkus is truly heaven's gift to software engineering. And I say that as a totally independent and unbiased observer.

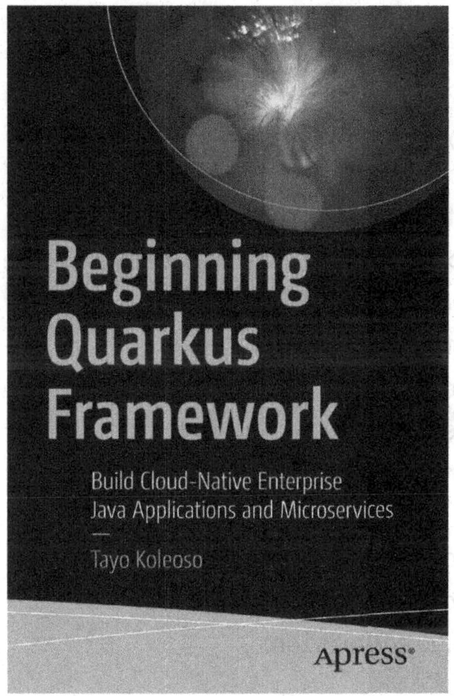

Yup. Totally unbiased

So, what can jOOQ do for you in Quarkus? Just like with Spring Boot, jOOQ *could* be all the SQL data access you need. It can also work with existing APIs in Quarkus like

- JPA
- Hibernate
- Reactive SQL
- SQL ResultSet streaming

The only thing that you can't do as of this writing is use the `@Query` annotation in `native` mode. Quarkus supports most of Spring Data JPA, except this bit. So, how does one get jOOQ into Quarkus?

Start with the Quarkus jOOQ extension:

```
<dependencyManagement>
   <dependencies>
      <dependency>
        <groupId>io.quarkus</groupId>
        <artifactId>quarkus-bom</artifactId>
        <version>${quarkus.platform.version}</version>
        <type>pom</type>
        <scope>import</scope>
      </dependency>
    </dependencies>
  </dependencyManagement>
<dependency>
      <groupId>io.quarkiverse.jooq</groupId>
      <artifactId>quarkus-jooq</artifactId>
      <version>0.2.2</version>
</dependency>
```

The preceding snippet is an excerpt of what you'll need to add to your Maven POM (or the Gradle equivalent) to add the jOOQ extension to your Quarkus, in addition to standard jOOQ dependencies. This is not necessarily the best way to add extensions to quarkus. Ideally, you would use the Quarkus maven plugin like so:

```
mvn quarkus:add-extension -Dextension=quarkus-jooq
```

That's the canonical way of installing Quarkus extensions. It doesn't work for the jOOQ extension because it's not (yet) in the registry of Quarkus extensions. The jOOQ extension is not an "official" Quarkus extension in that it's not built and maintained by the Quarkus core team. It's part of what they call the Quarkiverse,[3] an extended extension ecosystem that gives ownership to the developer community. The Quarkus-jOOQ extension is the hard work[4] of individuals outside the Quarkus team, backed by popular demand. Shout out to the gang!

Next up, install the appropriate JDBC extension for your database of choice:

```
mvn quarkus:add-extension -Dextension=jdbc-mysql
```

With the Quarkus extensions installed, you then configure your datasource settings in the `application.properties` file of your Quarkus app:

```
quarkus.datasource.db-kind=mysql
quarkus.datasource.username=dbuser
quarkus.datasource.password=thisisaterriblepassword
quarkus.datasource.jdbc.url= jdbc:mysql://localhost/edens_car
quarkus.jooq.dialect=mysql
```

[3] https://github.com/quarkiverse/quarkiverse/wiki

[4] https://github.com/quarkiverse/quarkus-jooq

Configuring the dialect is mandatory for any of this to work at all. With this basic config in place, you can get a DSLContext anywhere in the application:

```
@Inject
DSLContext dslContext;
```

The @Inject annotation is the Context and Dependency Injection (CDI) equivalent of the @Autowired annotation from the Spring framework. Quarkus supports both for the same purpose. The quarkus-jooq extension ships only with the community edition of jOOQ. You can override the community edition with the pro version if you have it inside your POM.xml.

Now, let's talk about packaging and testing all this stuff.

CHAPTER 5

Packaging and Testing jOOQ

Home stretch! Let's wrap the jOOQ roadshow up by talking about how jOOQ can fit into "modern" software development motifs like Continuous Integration/Continuous Development (CI/CD), containers (Docker, Podman, etc.), etc.

But first, let's level-set a little bit before we get into the meat (or vegetables if you like) of things:

- jOOQ allows you to Bring-Your-Own-SQL (BYOS).
- jOOQ will generate code for you, code that you most likely need at compile time.
- Your generated code becomes part of your business logic to do...stuff.

In today's world, you'll need to

- Be able to validate that your custom SQL works – either SQL you wrote yourself or packaged by another developer or team and handed to you.
- Be able to manage incremental changes made to your data model – changes originated either by yourself or another part of your organization. How would you support a new table or column added to the data model?

T. Koleoso, *Beginning jOOQ*, https://doi.org/10.1007/978-1-4842-7431-6_5

- Make practical and effective decisions about how and where your generated code lives. It's not uncommon (and maybe even preferred) that your entities and DTOs are packaged in a separate JAR file and included as a dependency in multiple software projects.
- Run integration tests without needing a whole standalone database server available at the point of testing. Take a Jenkins build server, for example: it's ideal that your build job doesn't need a standing MySQL server for your integration tests to run.

So, what are your options when you need to...

Package Code with jOOQ

This is going to be Maven-centric, sorry.

We've seen how to generate jOOQ code using the jOOQ Maven plugin, programmatically and from the command line. What we haven't talked about is *where* to put the generated code.

From the Maven standpoint, `src/target/generated-sources` is the recommended home for generated code, whether it's by jOOQ or anything else. Assuming you have the `jooq-codegen` plugin configured in your POM.xml like I demonstrated in Chapter 3, running `mvn package -DskipTests=true` will

- Connect to the database as configured in `jooq-configuration.xml`
- Generate the necessary code
- Compile the entire kit
- Skip running the tests
- Generate a JAR file in the `target` directory

Let's consider a couple of scenarios where you might want to deviate from this path a little.

When You Don't Need Code Generation

Code generation is all fine and dandy, but sometimes, you just want to build your kit and skip the code generation bit. Maybe you've already generated the code once, and nothing has changed; or you have a large schema you'd rather not deal with right now; or you're running the build in an environment that won't support the code generator. Configure a Maven profile like so:

```
<profiles>
        <profile>
            <id>no-jooqing</id> (1)
            <build>
                <plugins>
                    <plugin>
                        <groupId>org.jooq</groupId>
                        <artifactId>jooq-codegen-maven
                        </artifactId>
                        <version>3.15.1</version>
                        <executions>
                            <execution>
                                <id>jooq-codegen</id>
                                <phase>none</phase> (2)
                            </execution>
                        </executions>
                    </plugin>
                </plugins>
            </build>
        </profile>
    </profiles>
```

This is a standard Maven profile configuration you can add in at any point in your POM as a top-level element. What I've done here is

1. Configure a profile named `no-jooqing`. Inside this profile, I've defined the basics of the jOOQ code generator plugin. This configuration snippet mirrors the configuration of the same plugin in the `build` section of the POM. The idea here is for this profile-bound definition to override the other main plugin definition.
2. I set its phase execution to none, meaning that this plugin should not kick in at any point.

With this setup, I can then run a maven build thus:

```
mvn package -DskipTests=true -P no-jooqing
```

The `-P` flag activates my profile by the name of `no-jooqing`, thereby suppressing the code generator plugin. Arguably, there are simpler ways to achieve this effect, but profiles provide the most comprehensive way to selectively execute plugins. For example, you could choose to run a different jOOQ generator configuration based on, say, the JDK that's installed in the build environment:

```
<profile>
    <activation>
      <jdk>14</jdk>
    </activation>
    ...
  </profile>
```

With the preceding snippet, I've configured my profile to kick in only when the build is running on JDK 14 – the first version of the JDK to provide official support for the Records API (which jOOQ can generate as POJOs). You can have profiles activated based on operating system environment variables and other conditions. It's truly the most powerful option. Pretty cool huh?

When You Don't Have an Active Database Connection

It happens: you want to generate jOOQ code, but you don't have access to the underlying database server to connect to at build time. But thankfully, you have the Data Definition Language (DDL) that describes the schema. jOOQ provides the `org.jooq.meta.extensions.ddl.DDLDatabase` generator component, so you can generate code straight from a `.sql` script. Check it out:

```
<generator>
    <database>
        <name>org.jooq.meta.extensions.ddl.DDLDatabase</name>
        <properties>
            <property>
            <key>scripts</key>
            <value>src/main/resources/db-dump.sql</value>
        </property>
       ...
   <database>
<generator>
```

The `scripts` property accepts a path to the DDL script that will be loaded for code generation. This way, you're not shackled to a database server at build time. I should mention that this feature isn't restricted to just packaging usage - you can use this in any scenario it fits in.

Pro Tip Use `-- [jooq ignore start]` and `-- [jooq ignore stop]` to wrap SQL that should be ignored in your DDL script. What this means is that if your DDL contains `-- [jooq ignore start] CREATE TABLE ignore_me_please ... -- [jooq ignore stop] CREATE TABLE business_as_usual ...`, `CREATE TABLE ignore_me_please` will be ignored by `DDLDatabase`.

When Your Schema Needs to Incrementally Evolve

Have you heard of the evolutionary database pattern?[1] Its fundamental thesis is this: make incremental changes to your database schema, just like you already do with code. Whether you're starting with a fresh, empty database or you have an existing schema, you'll typically have a tool that

- Is able to apply new changes to a data model (DDL) or raw data (DML)
- Keeps a history of changes that have been applied, providing room to roll back incompatible or breaking changes
- Supports versioning of changes applied to a database
- Helps your code stay aligned with the database that it depends on

The two biggest players in this space right now are

- Flyway (`www.flywaydb.org`)
- Liquibase (`www.liquibase.com`)

[1] `https://en.wikipedia.org/wiki/Evolutionary_database_design`

They both operate on the same fundamental premise:

1. Provide your database changes in an agreeable file format, along with relevant versioning information.
2. They will apply your database changes to whichever database you point them.

How does jOOQ factor into all of this? Well, more so than your Hibernates and JPAs, jOOQ is fairly tightly coupled to the state of your data schema vis-à-vis code generation. The last thing you want is for your generated code to be referring to a trigger or function that no longer exists.

jOOQ has native support for Liquibase by way of the `org.jooq.meta.extensions.liquibase.LiquibaseDatabase` and the `jooq-meta-extensions-liquibase` Maven artifact. I'm a Flyway man myself, largely because Flyway doesn't require a specialized configuration syntax or DSL; and also, I'm a strategically lazy person.

For basic Flyway usage, simply provide your `.sql` file with a version format similar to the following:

`V1__Your_Descriptive_File_Name_Here.SQL`

The `V1` part of the filename is key. Subsequent updates to the schema should increment the version number to support the incremental change mechanism with Flyway. Keep all the SQL files in `/src/main/resources/db/migration` and you're ready to roll. At this point, you should add the Flyway dependency to your POM.xml:

```
<dependency>
       <groupId>org.flywaydb</groupId>
       <artifactId>flyway-core</artifactId>
       <version>7.14.0</version>
 </dependency>
```

With that in place, you're ready to run Flyway. You have the option of a command-line approach, a containerized approach (more on that later), or programmatic one. Let's have a look at the programmatic approach:

```
Flyway flyway = Flyway.configure().dataSource(jdbcUrl,dbUser ,
dbPassword).load();
flyway.migrate();
```

It's really that simple. Flyway will look for the latest versioned SQL script in `/db/migration` and apply the changes to the database you point it to. It also takes the previous versions of your schema into account, so that when you have up to a `V10__my_schema_update.sql`, the changes up till that point are taken into account. It supports baselining your migrations so you could choose to baseline your schema at, say, `V7__new_db_baseline.sql`, and that'll be where it starts considering migrations from.

Everything I've spoken about Flyway up till this point is super configurable by the way; I'm sticking with the defaults here for the purpose of demonstration. Because Flyway isn't necessarily the point of this section. No, here, I want us to consider how an evolutionary database model can support the goals of jOOQ to help generate and package the most current code based on an up-to-date but evolving schema. It gets even trickier when one is considering running this in a CI/CD, DevOps-heavy environment. You can't count on always having a standing database server connected to your Jenkins host, for Flyway or jOOQ to run against.

The ideal setup is a self-*contained* software project that can run its own code generation inside of itself at any point in the project lifecycle. Be it on the developer's machine, during a pre-merge step in the code repository, or as part of a build pipeline. No need to make sure some database server is up. No need to be worried about ruining the schema or database for another developer making concurrent changes to the same database. Yeah, that would be pretty sweet, wouldn't it?

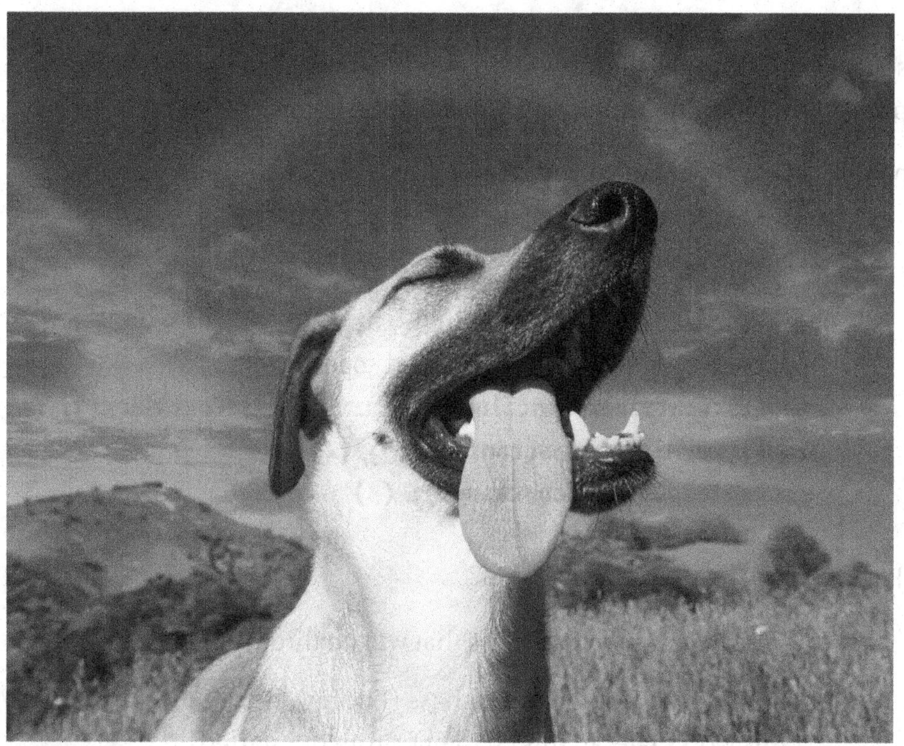

Blissful

One way to achieve this nirvana of self-sufficiency in your code is with a toolkit called TestContainers. I talk about TestContainers in more detail a little later in this chapter. For now, suffice it to say that TestContainers is arguably the best way to always have a full-strength database available and bundled with your application.

Recipe for a Self-Sufficient Database Project

Disclaimer: this is a hack. In the absence of, say, a dedicated TestContainers maven plugin, you'll need to get creative to be able to run a TestContainers-supported project outside of the testing phase.

But before we see how TestContainers can deliver a truly self-sustaining application, we should look at how I can package my project to support my ambitions. Consider the following class:

```
public class PreflightOperations {
    final static Logger logger = LoggerFactory.
    getLogger(PreflightOperations.class);

    public static void main(String[] args){
       logger.info("Running preflight operations");
       GenericContainer container = startDatabaseContainer(); (1)
       runFlywayMigrations(container); (2)
       generateJooqCode(container); (3)
    }
}
```

This is almost a vanilla Java class that will do three things:

1. Start a TestContainers database (I'll show how this works later in this chapter). From the started database container, I'll be able to obtain a database connection.

2. Using the obtainer database connection, I should be able to run my Flyway migrations immediately after.

3. Once my schema updates have been applied to my database container, I can then run the jOOQ code generator programmatically.

Pretty straightforward right? The question now is: how can I have this custom code run as part of a build process? Maven? Maven:

```
<plugin>
        <artifactId>maven-compiler-plugin</artifactId>
         <version>3.8.1</version>
```

```
        <executions>
            <execution>
                <id>pre-compile</id>
                <phase>generate-sources</phase> (1)
                <goals>
                    <goal>compile</goal>
                </goals>
            </execution>
        </executions>
    </plugin>
    <plugin>
        <groupId>org.codehaus.mojo</groupId>
        <artifactId>exec-maven-plugin</artifactId>
        <version>3.0.0</version>
        <executions>
            <execution>
                <phase>process-sources</phase> (2)
                <goals>
                    <goal>java</goal>
                </goals>
            </execution>
        </executions>
        <configuration>
            <executable>java</executable>
            <mainClass>com.apress.jooq.generator.
            PreflightOperations</mainClass> (2a)
           <cleanupDaemonThreads>false
           </cleanupDaemonThreads>
        </configuration>
    </plugin>
```

The capability to run java code as part of the build process rests on two Maven plugins:

- maven-compiler

  This plugin will compile source code. Because my `PreflightOperations` class is still raw source code, I need to compile it before I'm able to run it as part of the build process.

- maven-exec

  This plugin will run any arbitrary executable. Choosing `java` for the `executable` parameter preps the plugin to execute a Java class with a `main` method.

Together, these plugins sing beautiful music:

1. maven-compiler starts by compiling my source code in the `generate-sources` phase of the Maven build process. This is going to ensure that I have a compiled `PreflightOperations` class to run with...

2. maven-exec, the plugin that allows me to run arbitrary executables. I've chosen to run this plugin in the `process-sources` phase, which comes immediately after the `generate-sources` phase. It's at this point that the database will start, my Flyway migration will execute, and then jOOQ gets to generate any necessary new source code.

   a. I provide the fully qualified class name (FQCN) to my class that does the execution.

b. Because TestContainers does a lot of background processing on daemon threads, there's the risk that it won't be ready to quit by the time the maven-exec plugin is ready to move on. `cleanupDaemonThreads` allows the build process to proceed, while TestContainers does its thing.

Easy peasy. I want to reiterate: this is a hack. The gold standard should be sans custom code and configuration only. Additionally, the generated code/entities in general could benefit from a lot more modularity.

The only thing missing from this recipe is the actual dynamically instantiated database itself. We'll learn about that when we talk about...

Testing with jOOQ

Also known as: sleep well at night. I'm an absolute nutter for automated testing, especially integration testing.

Uh oh. I just used a buzzword: "integration testing." Integration testing tends to be conflated with a bunch of other things that (in my layman's opinion) don't qualify. Allow me to pontificate.

I believe the industry has settled on the scope of unit testing, that is, validating the behavior of standalone units of code, for example, a function or a method. You're not concerned about how these functional units interact together to deliver a business scenario. You're likely going to mock out every single dependency that the method under test has, to focus solely on what's within the curly braces.

Then we have "end-to-end" testing, where you're crossing multiple system boundaries - front-end to back-end integration tier, etc. This is what some folks call "QA" testing - making sure *everything* works together to satisfy the user's needs.

Somewhere in the middle of unit and end-to-end testing is where you find integration testing and the sometimes *testy* debate around what it actually means.

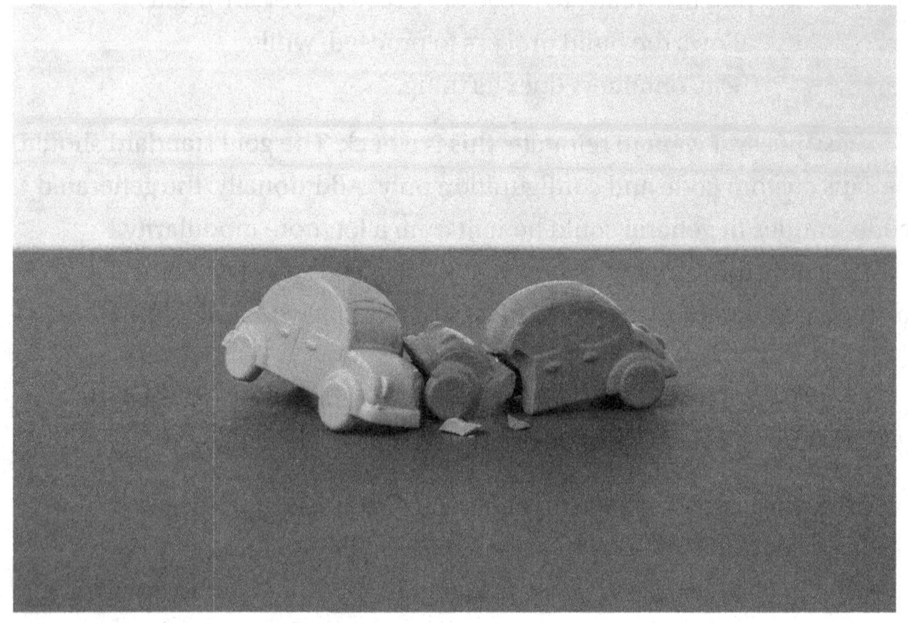

Pictured: Integration testing. Probably.

For the purposes of this section, integration testing is how you make sure that carefully selected slices of your code work well together. In a typical integration test, you'd want to string a handful of components together and see if they all behave the way you expect. Ideally, your integration tests are closely aligned with use cases that the business/users expect.[2]

This is not to say jOOQ doesn't have tooling to support unit testing – far from it. I'm just personally more invested in integration tests that give me confidence about the product I'm putting out (vs. vanity metrics

[2] `www.agilealliance.org/glossary/bdd/`

around testing). Just ask Kent Beck[3] how he feels about writing tests for tests' own sake.

Based on the features that jOOQ provides, what is there to test anyway?

- The syntactic correctness of plaintext SQL statements that *you* bring to jOOQ. jOOQ's own SQL is highly unlikely to be incorrect.
- The semantic correctness of both yours and jOOQ's own SQL. jOOQ protects you from writing syntactically incorrect SQL. It's still a solid idea to validate the semantic correctness of the SQL, generated or otherwise.
- The accurate reflection of your generated code vis-à-vis the database schema. You're not going to have a good time if your generated code is even a little bit out of sync with the underlying schema for whatever reason.

Tools of the (SQL) Testing Trade

Let's see what's at our disposal for testing SQL. To be clear, these are not solely for SQL testing, but y'know, this is a SQL-in-Java book, sooo...

1. JUnit (`www.junit.org`)

   The premier testing in Java – all the others are posers.[4] JUnit 5 (codename Jupiter) is your one-stop shop for all your testing needs. The latest version supports pretty much every testing paradigm you could imagine: Behavior-Driven Development (BDD), Acceptance Test-Driven Development

[3] `https://stackoverflow.com/a/153565/1530938`

[4] Editor's note: Hot take!

(ATDD), unit and integration tests, etc. It ships with a suite of annotations that provide all manner of convenience for testing your Java code at varying levels of granularity. But you probably have already heard of it.

2. Mocking Frameworks

   A mocking framework (e.g., Mockito, PowerMock) will help you stub out different parts of your code – hardly a new concept, I know. Stubbing or mocking selected sections of your code while testing allows you to laser-focus your tests to only what matters to you. Where things could possibly get spicy is having to work with some of jOOQ's static methods. Hang tight while we dig into those a bit more.

3. Embedded Databases

   In the course of testing, you'll eventually need to be able to dynamically

   - Load a schema into a database on demand
   - Load/destroy data in a database on demand
   - Sequentially run test methods that depend on shared state as part of a test scenario

   All of these scenarios require that your software project have a database ready quickly and flexibly. That's where the embedded or "in-memory" databases come in. They're databases that are designed for dynamic and flexible usage in lightweight scenarios, for example, testing. Examples of these include

- H2 (`https://h2database.com/html/main.html`)
- HSQLDB (`http://hsqldb.org/`)
- Derby (`https://db.apache.org/derby/`)

Yes, they're all written in Java. With these, you can have a database "server" available at any point in your development lifecycle without needing an actual DB server deployed anywhere.

Now because they're lightweight, their capabilities are limited. So, you'll typically be missing some fundamental features. Things like check constraints, triggers, even the `LIMIT` SQL keyword may not be supported depending on which vendor you choose. They're lightweight for good reason: fast, highly efficient database operations without "frills." If you prefer a full-strength *and portable* database for your testing, you should turn to...

4. Containerized Databases

   You can get most full-strength databases like MySQL, PostgreSQL, and Oracle in a containerized format compatible with Docker, Podman, and other container runtimes. What are containers? I'll get into more detail when we get to that point, but for now suffice it to say this: containers are portable versions of your favorite software packaged in what's known as images. These portable packages will typically *contain* complete operating system installations with all the trimmings; your desired software can then be bundled with these complete OSes and delivered via a centralized registry. Containerized databases (mostly) deliver the full strength of your preferred database server while keeping them portable enough to start up an instance programmatically/dynamically. This way, you can have full-strength databases available whenever you want them, for example, as part of a pipeline, a build script, or a JUnit integration test. No half-assing it.

5. jOOQ's Testing Kit

   There are a few components in the jOOQ toolkit that support your testing and validation needs. Check it out:

   - `org.jooq.tools.jdbc.MockConnection`, `org.jooq.tools.jdbc.MockDataProvider`, and a couple of related `Mock*` components help to mock out different parts of the query operations in jOOQ.
   - `org.jooq.Parser` can be used to validate your SQL queries by attempting to produce jOOQ artifacts out of your plaintext SQL.

Different combinations of the preceding tools will give you the peace of mind you need while programming with jOOQ. Not to mention the various testing facilities offered by ecosystems like the Spring framework and Quarkus – there's a bunch of powerful testing techniques in both of them.[5] Ultimately, what I would want in my project is a self-contained, self-sufficient kit that can run its own tests anywhere, without too much of a dependence on its operating environment. This portability becomes more crucial when you're operating in a CI/CD environment. Let's see how all of these play together.

[5] I'm particularly fond of the powerful `QuarkusUnitTest` class that isn't publicly documented (but it's in my book). Highly recommend for integration tests.

Unit Testing with jOOQ

Consider the following jOOQ query method:

```
public static void selectWithOptionalCondition(boole
an hasFilter, Map<String, Object> filterValues) throws
SQLException {
        try (Connection connection = DriverManager.
        getConnection("jdbc:mysql://localhost/edens_
        car?user=root&password=admin")) {
            DSLContext context = DSL.using(connection,
            SQLDialect.MYSQL);
            Condition conditionChainStub = DSL.noCondition();
            if (hasFilter) {
                for(String key: filterValues.keySet()){
                    conditionChainStub = conditionChainStub.
                    and(field(key).eq(filterValues.get(key)));
                }
            }
            List<CompleteVehicleRecord> allVehicles =
            context.select().from(table("complete_
            car_listing")).where(conditionChainStub).
            fetchInto(CompleteVehicleRecord.class);
            logger.info(allVehicles.toString());
        }
    }
```

I'm doing some fancy construction for the WHERE clause of the jOOQ query, dynamically constructing the Conditions that would be translated into that clause. Beside the database query that ultimately needs to happen, how can I validate that my condition chaining is going to result in the WHERE clause I expect? That's where a unit test comes in.

Using Mockito

Mockito is a pretty popular mocking framework like I mentioned earlier that allows you to stub out parts of your code that don't need to be invoked during testing. It also allows you to substitute parts of your code with something else to facilitate specific testing scenarios. For my use case, I want to validate that the condition chaining in my query is working correctly – I don't need query results for that. I'm going to start by adding Mockito as a dependency to my project:

```
<dependency>
            <groupId>org.mockito</groupId>
            <artifactId>mockito-inline</artifactId>
            <version>3.12.1</version>
        </dependency>
 <dependency>
           <groupId>org.mockito</groupId>
           <artifactId>mockito-junit-jupiter</artifactId>
           <version>3.12.1</version>
  </dependency>
```

These Maven dependencies will furnish my project with the necessary libraries to use Mockito. The `Mockito-inline` artifact is especially crucial because it provides the support for mocking static methods. The need for that feature will become apparent shortly. The `mockito-junit-jupiter` artifact is prescribed for the latest version of JUnit; for older versions of JUnit, use `mockito-core` instead. On to the code!

```
@ExtendWith(MockitoExtension.class) (1)
@TestInstance(TestInstance.Lifecycle.PER_CLASS) (2)
@DisplayNameGeneration(DisplayNameGenerator.ReplaceUnderscores.
class) (3)
class JooqUnitTests {
```

```
    static MockedStatic mockedDriver; (4)
    final Logger logger = LoggerFactory.getLogger
    (JooqUnitTests.class);

    @BeforeAll (5)
    public static void prepare(){
           mockedDriver = mockStatic(DriverManager.class); (5b)
    }
 //more to come
}
```

The preceding snippet demonstrates the setup of some test fixtures I'm going to need shortly:

1. `@ExtendWith` is a JUnit component that allows one to plug in to the runtime with custom code. Different vendors can then supply a class that'll fulfill the contract and be usable here. In this case, I'm using Mockito's `MockitoExtension` class. That introduces Mockito's features into this test unit.

2. `@TestInstance` is a JUnit component that configures the lifecycle of the test class. With `Lifecycle.PER_CLASS`, I've specified that I want a single instance of `JooqUnitTests` to be reused for any number of test methods inside the class. This way, the test methods can share state across multiple invocations.

3. `@DisplayNameGeneration` determines how the test cases will be displayed in reports, your IDE, and elsewhere. With `ReplaceUnderscores`, I can use underscores in my test method names and they'll be replaced with spaces during display. This way,

the method names can be user-friendly sentences that even non-engineers (e.g., Product Owners) can understand and consume.

4. `MockedStatic` is another Mockito test fixture that allows me to mock static methods and interfaces. I'm going to be using it to stub the `DriverManager.getConnection` interaction from JDBC.

5. `@BeforeAll` stipulates that the annotated method – `prepare` – be run **once** before any test methods are run...

    a. So that I can customize the behavior of the `DriverManager` class to suit my needs

With the prep work out of the way, let's crack on with the unit test proper. Hang on to your seat, it's a lot of typin':

```
@ParameterizedTest (1)
@CsvSource({                    (1a)
        "BLUE,2020",
        "SILVER,2020"
})
void test_dynamic_condition_api(String color,String year)
throws SQLException {
        MockDataProvider mockJooqProvider = context -> { (2)
                        MockResult[] results = new
                        MockResult[1];
                        String sql = ctx.sql();
                        logger.info(()->"Binding 1: "+ctx.
                        bindings()[0]);
                        assertAll(()->{
```

```
                    assertTrue(ctx.bindings().
                    length == 2 ); // validate two
                    parameters are bound;
                    assertEquals(ctx.bindings()
                    [0],color);
                    assertEquals(ctx.bindings()
                    [1],year);
                });
                CompleteCarListingRecord
                completeCarListing = new
                CompleteCarListingRecord();
                results[0] = new MockResult(completeCar
                Listing);
                return new MockResult[0];
            }
        };
        MockConnection mockConnection = new MockConnection(
        mockJooqProvider); (3)
        mockedDriver.when(()-> DriverManager.
        getConnection(anyString())).
        thenReturn(mockConnection); (4)
    JooqApplication.selectWithOptionalCondition(true,Map.of
    ("color",color,"year",year));
    }
```

The primary goal of this test is to make sure that filter parameters are processed correctly. As a secondary goal, I don't want or need an actual execution of the query against a database. So I need to substitute the JDBC `Connection` usage for something else. That's where jOOQ's `MockConnection` and `MockDataProvider` come in:

1. JUnit provides `@ParameterizedTest` allowing us to feed data into a test method from multiple sources.

    a. Here, I'm using the `@CsvSource` option to simulate CSV data being passed in. For every row I provide, JUnit will parse the columns and feed them to the test method as method arguments.

2. To provide a `MockConnection` from jOOQ that will replace a legit JDBC connection, I need to build out a `MockDataProvider`.

    a. Inside my implementation of `MockDataProvider`, I have access to some pretty nice test fixtures like a `MockExecutionContext`, the SQL that's going to be executed and crucially: the parameter bindings supplied to the query. These I then validate to ensure they're present and the right count. There's a lot of flexibility in here to allow many testing use cases.

    b. The contract for the `MockDataProvider#execute` requires me to return an array of `MockResults`. Since I don't really care about the result in this scenario, I just construct an empty `Record` from a generated class and move on.

3. Having implemented my `MockDataProvider`, I can go ahead and construct a `MockConnection`.

4. Remember earlier when I mocked out `DriverManager` with `MockedStatic`? Well now's its time to shine! Having stubbed out `DriverManager`,

I can stipulate that when any string is passed to the `getConnection` method, my `MockConnection` should be returned instead of an actual JDBC connection.

With all that setup, I can then execute my business logic and see how things shape out. No data will be retrieved; it's all isolated to that one method.

With SQL Parsing

jOOQ ships with some à la carte SQL parsing capabilities that don't necessarily have anything to do with SQL execution. You can use the `Parser` class to generate jOOQ components from plaintext SQL; in the process, it'll let you know whether your SQL's legit or not. Observe:

```
@Test
void validate_my_dodgy_sql(){
      assertThrows(ParserException.class, ()->
                  DSL.using(SQLDialect.MYSQL)
                         .parser()
                         .parse("selecet * from table group by
                         1 where having max (column) > 10"));
}
```

That SQL ain't right,[6] I'm sure you'll agree. With `assertThrows` from JUnit, I've specified that I expect this attempt to `parse` the plaintext SQL should fail with a `ParserException`. *Via con Dios!*

[6] Yes, I'm a huuuge King of the Hill nerd. You should be too.

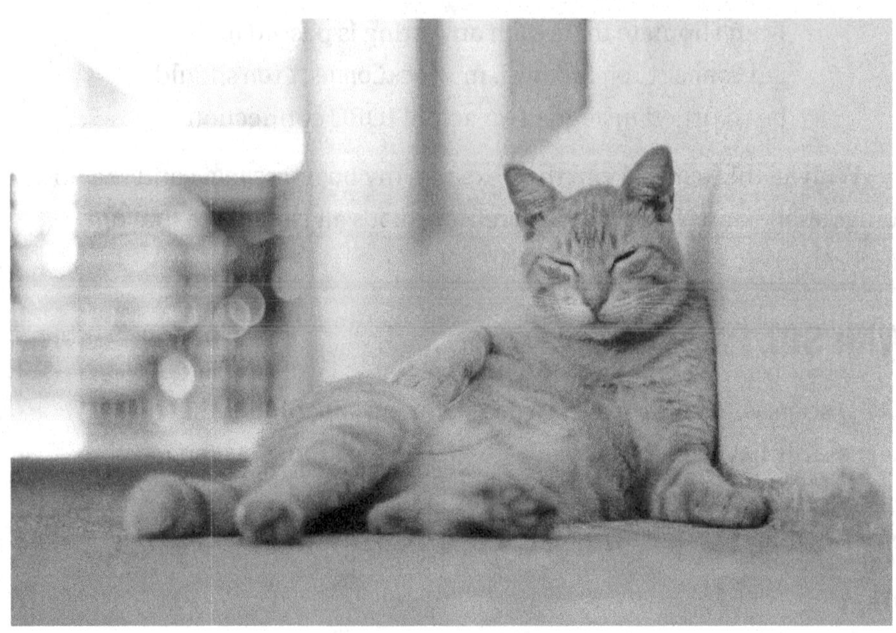

Yawwwwwn

Trust me, guys: BDD is where it's at. When you're working with data, you really want to get your hands dirty with actual executions of SQL statements; see real results. Are you with me? I hope so. Because we're about to get to the fun part!

Integration Testing with Docker and TestContainers

Docker, like I mentioned a couple of pages ago, is a runtime for containers. If you're unfamiliar with the concept, think of Docker like a virtual machine – a JVM if you will. Just like you can download any kind of JAR that's packaged by a third party, and run inside your JVM, Docker functions similarly. Different vendors publish images to Docker Hub;[7]

[7] https://hub.docker.com/

you can then pull these images down and run containers based on the image. In a sense, Docker Hub is the Maven Central of the container world. You can get almost any major piece of software as an image and, thus, a container. This gives you a ton of flexibility and portability, allowing you to run previously bulky and overhead-intensive software in a portable and mostly lightweight format, so that you can run entire operating systems, CI/CD servers and tools, critical infrastructure software, and, yes, databases, in a dynamic and flexible form factor.

TestContainers is a Java library that helps you take the portability of containers further. It gives you the *power* to run any containerized software from within Java code.

Testcontainers

`www.testcontainers.org`

We shall now use TestContainers to start up a MySQL database server as part of our JUnit tests, as well as load it up with real tables and data. Then, we can run actual code against it – none of that mocking business here. Well, maybe just a little. Let's go!

First, download/install Docker for your operating system – www.docker.com is a good place to start for most users. TestContainers depends on the Docker runtime to do its magic. Can't run a JAR file without a JVM, can you?

As usual, the Maven dependencies come first:

```
<dependency>
        <groupId>org.testcontainers</groupId>
        <artifactId>testcontainers</artifactId>
        <version>1.16.0</version>
        <scope>test</scope>
    </dependency>
    <dependency>
        <groupId>org.testcontainers</groupId>
        <artifactId>junit-jupiter</artifactId>
        <version>1.16.0</version>
        <scope>test</scope>
    </dependency>
    <dependency>
        <groupId>org.testcontainers</groupId>
        <artifactId>mysql</artifactId>
        <version>1.16.0</version>
        <scope>test</scope>
    </dependency>
```

Like with Mockito, please pay attention to the artifacts. For each database version it supports (and there are many), TestContainers has a dedicated Maven dependency. For my use of MySQL, I've added the mysql artifact; choose correctly for your database container of choice.

Now, for my next demonstration, a little bit of test prep:

```
@TestInstance(TestInstance.Lifecycle.PER_CLASS)
@DisplayNameGeneration(DisplayNameGenerator.ReplaceUnderscores.
class)
@Testcontainers (1)
@TestMethodOrder(MethodOrderer.OrderAnnotation.class)
public class JooqIntegrationTests {

    static MockedStatic mockedDriver;
    final static String DATABASE_NAME = "edens_car";
    final static String USERNAME = "auserhasnoname";
    final static String PW = "anawfulpassword";

    @Container (2)
    static GenericContainer mySqlContainer = new
    MySQLContainer(DockerImageName.parse("mysql:latest")) (3)
            .withDatabaseName(DATABASE_NAME)
            .withInitScript("schema_with_data.sql")
            .withUsername(USERNAME)
            .withPassword(PW)
            .withEnv("TESTCONTAINERS_CHECKS_DISABLE","true")
            .withExposedPorts(3306);

   @BeforeAll
   public static void setup(){
      mockedDriver = mockStatic(DriverManager.class); (4)
   }
}
```

Now remember, this is all just prep inside a test class; the actual test methods will follow shortly:

1. `@TestContainers` is a JUnit extension provided by the TestContainers library. It's really a façade for the `ExtendWith` annotation we saw earlier with Mockito.

2. `@Container` is also supplied by TestContainers. With this annotation, TestContainers can hook into the lifecycle of the JUnit runtime and prepare the container instance ahead of time.

3. The `GenericContainer` class is a...generic class that wraps most of TestContainers' container-based functionality.

    a. Just like we've had to do with Maven and any other dependency management system, I have to supply named coordinates to the appropriate Docker image in the hub. The format is "name":"tag". Here, I'm saying I want the "mysql" image with the "latest" tag or version.

    b. `withDatabaseName` allows me to set a name for my yet to be created database.

    c. `withInitScript` defines the name of a SQL script file that will be loaded immediately after the container has completed initialization. This allows me to populate my database with DDL and DML in advance of any actual test execution.

    d. I seed the database container with credential info `withUsername` and `withPassword`.

    e. I can also deliver random environment variables to my container with `withEnv`. Here, I'm supplying a TestContainers command-line parameter that allows it to skip some startup checks, thereby getting the database container ready, faster.

    f. Finally, I define the port on which MySQL should be listening. Note that this is still internal to the container. A separate, random port will be published by TestContainers for me to be able to connect to the MySQL container. This process is called port mapping in the container world.

4. Finally, like we did before, I prepare to mock out the `DriverManager` because I want to provide a dynamically generated `Connection` – but a real one this time to a real MySQL database.

Now that I've set up all my test fixtures, I can go ahead and write the actual test:

```
@Test
 public void test_containerized _connection() throws
SQLException {
        JdbcDatabaseContainer container =
        (JdbcDatabaseContainer) mySqlContainer; (1)
        Connection connection =container.
        createConnection("");    (2)
        mockedDriver.when(()-> DriverManager.
        getConnection(anyString())).thenReturn(connection); (3)
        JooqDemoApplication.insertVehicle();
    }
```

In the preceding snippet

1. I cast the `GenericContainer` to a more specialized form, the `JdbcDatabaseContainer`.

2. This now allows me to directly obtain an instance of a JDBC `Connection` right off the container.

3. I can then substitute my own `Connection` into the `DriverManager`.

Following all that, I can then directly execute my test logic code.

This is a **proper** test. It contains actual data, actual database fixtures and trimmings, all inside a real database. Delightful.

Here's another nifty thing JUnit allows: test method ordering. With test method ordering, you can have tests that depend on each other or at the very least must run in a specific order. Check it out:

```
@Test
@Order(1)
public void test_containerized_connection() throws
SQLException {
    ...
 }

@Test
@Order(2)
public void test_valid_db_insert() throws SQLException {
    if(!mySqlContainer.isRunning()){
        mySqlContainer.start();
    }
    JdbcDatabaseContainer container =
    (JdbcDatabaseContainer) mySqlContainer;
    container.getJdbcUrl();
    Connection connection =container.createConnection("");
    DSLContext context = DSL.using(connection, SQLDialect.
    MYSQL);
    List<CompleteVehicleRecord> allVehicles = context.
    select(field(name("brand")), field("model"),
    field("price")).from(table("complete_car_
```

```
        listing")).orderBy(field("year").asc(), two()).
        fetchInto(CompleteVehicleRecord.class);
        assertTrue(allVehicles.size() == 1);
    }
```

The @Order annotation allows me to stipulate that test_valid_db_insert should execute immediately after test_containerized_connection. Here's where things get a tad wonky.

See, TestContainers is wired to shut down a container immediately after the test method is done executing. The container isn't actually destroyed, but it's not running. This is what makes it necessary to take some precautions when reusing a container instance across test methods. In this scenario, I've inserted data with test_containerized_connection; I then want to validate the insert in test_valid_db_insert. I must check that the container is still up with isRunning; otherwise, the test fails. If the container isn't running, I can restart it with start. This is a pretty crude mechanism to support container reuse; you can get a lot fancier and maintainable with it.

Pro Tip TestContainers provides the ScriptUtils.runInitScript utility that helps execute arbitrary SQL scripts against a database container. This way, even after an initial load into the database, you can execute custom SQL at any point in your testing.

This is all well and good in a "legacy is cute" sort of way. If you're using containers in your code, you're likely not dealing with DriverManager. You're likely a framework sort of person. How about we try this on...

With Spring Boot

Spring Boot is, well, Spring Boot. It provides a whole arsenal of test fixtures and componentry that could make one dizzy. We're not going to dig into all of that. We're just here for the jOOQy goodness. Check it out:

```
@SpringBootTest (1)
@Testcontainers
public class JooqSpringBootTests {

    @Autowired
    JooqBean jooqBean; (2)
     ...
    @Container
    static GenericContainer mySqlContainer = new
    MySQLContainer(DockerImageName.parse("mysql:latest"))
    ...
    @DynamicPropertySource (3)
    static void postgresqlProperties(DynamicPropertyRegistry
    registry) {
        JdbcDatabaseContainer container =
        (JdbcDatabaseContainer) mySqlContainer;
        registry.add("spring.datasource.url",
        container::getJdbcUrl);
        registry.add("spring.datasource.password",
        container::getPassword);
        registry.add("spring.datasource.username",
        container::getUsername);
    }

    @Test
    @Sql("/schema_with_data.sql") (3)
    public void test_springboot_loading(){
```

```
        List<Vehicle> vehicles = jooqBean.runSql();
        assertTrue(vehicles.size() >= 1);
    }
}
```

I've slimmed down this snippet to exclude old stuff you've seen up till now. We're here for the new and just the new:

1. With `@SpringBootTest`, Spring will take notice and make its facilities available.

2. This is how I can now inject my `JooqBean` containing all manner of jOOQ queries.

3. New with v2.2.6, Spring Boot provides the `@DynamicPropertySource` annotation which allows me to dynamically override any framework properties I choose. This comes particularly in handy when one is dynamically spinning up database containers of unknown port, username, and password.

4. Finally, on the test method itself, I deploy the `@Sql` component, also from Spring. This annotation will execute the SQL statements in the supplied script file. The default behavior is to execute the script *before* the test method is run, but that can be changed. Additionally, I can supply any number of scripts here for different purposes. Pretty neat.

Now that you've gotten yourself somewhat familiar with TestContainers, let's revisit our packaging dilemma: how can we apply changes to our schema, generate updated jOOQ classes, as well as run our tests all without needing an external database server? I've demonstrated some of the prep that needs to happen in support of this goal. Now let's see the code that'll back it up.

```
public static GenericContainer startDatabaseContainer() throws
SQLException {
         mySql = new MySQLContainer(DockerImageName.
         parse("mysql:latest"))
                .withDatabaseName(DATABASE_NAME)
                .withUsername(USERNAME)
                .withPassword(PW)
                .withEnv("TESTCONTAINERS_CHECKS_
                DISABLE","true")
                .withExposedPorts(3306);
        mySql.start();
        return container;
    }
```

The preceding snippet isn't too different from the code I've shown in the business of testing. The main difference here is that I'm explicitly starting the database container with the `start` method. Yes, there's a `stop` method as well for when you're done. After starting a containerized MySQL, I can then execute my migration with Flyway.

```
//run the migration with a connection to the database container
public static void runMigrations(GenericContainer container){
        JdbcDatabaseContainer container =
        (JdbcDatabaseContainer) container;
        Flyway flyway = Flyway.configure().
        dataSource(container.getJdbcUrl(),container.
        getUsername(),container.getPassword()).load();
        flyway.migrate();
    }
```

And then tying everything together:

```
public static void main(String[] args) throws SQLException {
            logger.info("Running preflight operations");
            GenericContainer mySql = startDatabaseContainer();
            runMigrations(mySql);
            generateJooqCode(mySql);
            connection.close();
            mySql.close();
    }
```

Thus, we can have a completely self-sufficient project, at least from the database perspective. This can be run on a developer machine or on a build server.

All told, you want a self-contained and self-sustaining software project kit that can

1. Portably evolve with a changing database schema
2. Run its tests wherever it lives – on a developer's machine, in a build pipeline, before a pull request merge, etc.
3. Validate your assumptions against production-like software and infrastructure without the associated overhead
4. Give you warranties as to the syntactical correctness of your database-related code

Because after all, this is what modern software development is all about.

Good luck, and thanks for reading!

Index

T. Koleoso, *Beginning jOOQ*, https://doi.org/10.1007/978-1-4842-7431-6

M, N

O

P

Q

R

S

T, U, V

W, X, Y, Z

Printed in the United States
by Baker & Taylor Publisher Services